Serpents in the Sand

Serpents in the Sand

Essays on the Nonlinear Nature of Politics and Human Destiny

COURTNEY BROWN

Ann Arbor

THE UNIVERSITY OF MICHIGAN PRESS

Published in the United States of America by
The University of Michigan Press
Manufactured in the United States of America
♾ Printed on acid-free paper

1998 1997 1996 1995 4 3 2 1

A CIP catalog record for this book is available from the British Library.

Library of Congress Cataloging-in-Publication Data

Brown, Courtney, 1952–
Serpents in the sand : essays on the nonlinear nature of politics and human destiny / Courtney Brown.
p. cm.
Includes bibliographical references and index.
ISBN 0-472-10643-0 (hc)
1. Political science. 2. Political science—Methodology. 3. Nonlinear theories. 4. Voting. I. Title.
JA74.B723 1995
320'.01'1—dc20 95-11859
CIP

For Azizi and Isabella

Contents

Preface

No one will ever criticize this book for not having a clearly defined message. Nonetheless, I suspect that the message itself will draw considerable attention, and perhaps controversy, from within the social scientific community. I alone, of course, am responsible for both the message—whatever its worth—as well as any errors that may be embedded within both the analyses and the fine print on the pages of this volume.

Some controversy may result as a side effect of my occasionally flamboyant prose, but of this I request my readers' tolerance. I do not hide the fact that I have something that I feel is important to say about the way we look at the politics of human society, and, indeed, about humanity itself. If I am to be judged due to what I say, let there be no ambiguity with regard to my point of view, and readers should remind themselves that the prose on these pages will remain the only defense of my ideas for years to come. In any court, be it one of law, public opinion, or the views of one's peers, no defense would be worth its salt if the advocate did not speak with sufficient potency to adequately make one's case, especially when the argument of the case runs against the current of an already entrenched way of thinking.

Some may feel that I am projecting a methodological discussion that is based on the preferences of individual researchers into the realm of the philosophy of existence. This is almost true. The part that is false is that the methodological orientation is not, in my view, a preference of individual researchers. Humans are either fundamentally and intrinsically nonlinear in their individual and collective behaviors or they are not. The question of the individual methodological preferences of researchers has nothing to do with our inherent nature. Individual researchers may argue about how to proceed with their investigations of a nonlinear universe, but their preferences can hardly make this universe linear if it is otherwise.

I am grateful to Emory University for its support to me during the course of my investigations. In particular, I was very significantly supported by the Information Technology Division at Emory. James Johnson, vice provost of Emory University's Information Technology Division, administered a grant to me that supported both my software and hardware needs for this research. Marie Matthews helped by leading me through the orchestration of a very diverse set of purchases relating to this grant. Moreover, these investigations consumed a large proportion of total mainframe computer usage at the university.

Some of the figures in this volume were created using imaging software provided free of charge by the National Center for Supercomputing Applications at the University of Illinois, Urbana–Champaign. This project has reinforced to me the importance of the center as a national, and indeed international, resource.

The Inter-university Consortium for Political and Social Research, another important international resource, supplied data for the analyses presented in chapters three, four, and five. I alone am responsible for all of the interpretations of these data.

I am particularly grateful to the Inter-university Consortium for Political and Social Research Summer Program. Hank Heitowit, the educational director of the program, invited me to teach (and proselytize) the application of nonlinear dynamical systems in social settings for 11 years. Nonlinearity is just now coming of age in the social sciences, and it is to Hank's credit that he both saw this as an inevitability so many years ago, and had the courage to argue for the continuation of this nonlinear component in the summer program's offerings since 1984. I am also grateful for the institutional support of Richard Rockwell, Clifford Clogg, and Carolyn Geda.

Certain individuals have helped me in a variety of ways. Some offered encouragement, others ideas, and yet others their time and energy. In particular, I am grateful to Chris Achen, Mike McBurnett, John Sprague, Robert Huckfeldt, Carol Kohfeld, Robert Boynton, Thad Brown, and Robert Stine.

My wife, Isabella, and my son, Azizi, helped in their own invaluable and unique ways.

CHAPTER 1

Nonlinear Politics

Very few naturally occurring phenomena in our universe evolve linearly. Galaxies spin in spiral arrangements, bullets travel in parabolic trajectories, the branches of trees spread in complex and curving patterns, and change in the ecosystem of fish ponds can be described in terms of nonlinear population dynamics. Indeed, wherever one looks, the behavior of nearly everything manifests itself in nonlinear ways. Why then do social scientists typically look at human behavior in linear terms? Why have we established and defended a large collection of computerized statistical methodologies that implicitly encourage a linear worldview of human existence? How could it be that in all of God's universe, we humans are the only linearly evolving creations?

The answer to these questions is that humans are no more linear than anything else in the universe. We use linear terms to describe our behavior for a variety of reasons, many of which are discussed below. But fundamentally, we have made a mistake in the way we view ourselves. We have an incorrectly prescribed set of lenses through which we look at human behavior. We are intrinsically nonlinear creatures, and the intellectual hegemony of linearity must end. Indeed, this book is written as a direct challenge to the linear paradigm. Within these pages, I argue the case for a nonlinear worldview with respect to the evolution of human political systems.

The Challenge of Nonlinearity

I have made a number of strong assertions in the above paragraphs. There are two ways to support these assertions. The first would be to examine the logic of each claim. This would require a more philosophical approach than I use in this book. Indeed, I would have to spend a great deal of time describing what other social scientists do while pointing out what could be accomplished if they did things differently. This is both a negative approach to the topic and a particularly unfruitful one as well. Moreover, it unfairly slights the value of current empirical social science that in fact has brought a tremendous degree of advancement to our understanding of human societies.

The approach that I use in this volume is to present the results of a variety of my own studies regarding nonlinear change in society. Thus, in part, this book is offered as a collection of nonlinear case studies. It is one thing to argue that nonlinearity should be a more integral aspect of social scientific research designs. It is quite another thing to conduct studies that do just that. But it would be a serious mistake to think that this book is a treatise about methodology, for it is not.

There are two basic aspects to my challenge to the linear worldview. The first rests at the level of social and methodological theory. In the pages that follow, I propose alternative nonlinear ways to look at the dynamics of human political behavior. These alternative perspectives necessarily have a mathematical basis, but this book is about politics not mathematics, and an effort has been made to make this book understandable to all individuals, regardless of mathematical background. The second aspect to my argument in favor of a nonlinear worldview of politics is associated with the use of concrete examples. The examples of nonlinear politics that are presented here not only serve as convenient heuristic vehicles with regard to matters of nonlinear model specifications, but they also serve the larger purpose of demonstrating how commonly nonlinearity can appear in a wide variety of political settings.

But, and again, this book is not a treatise on model building. The intellectual stakes are much greater than algebraic craftsmanship. Methodological practices are not the cause of the problems that social scientists have in viewing the world as either linear or nonlinear. Indeed, the algebra is the symptom of a much more serious intellectual condition. Ultimately, what is at stake is a philosophy of human social evolution, and this volume offers a view into what could be a new paradigm in social scientific thinking.

A Brief Recent History of Nonlinearity

Interestingly, nonlinearity in social scientific thinking is not new. While one can make a case for many points of potential origin, one of the most significant early contributions was made in mathematical sociology in the 1950s and 1960s. That period contained a number of thinkers who, in retrospect, seem to have been far ahead of their time. In particular, three people stand out as having made major contributions in thinking about nonlinearity in human society. These individuals are James S. Coleman (1964), William N. McPhee (1963), and Herbert A. Simon (1957). Unfortunately, their early efforts never attained the status of a movement. Indeed, after the surge in the use of statistical models based on general linear forms, it seemed as if nonlinearity was dead in the social sciences.

No science maintains its place in society without continuing advances. The problem with the early nonlinearists was that the application of nonlinear mathematics required computers that were faster than those available in the

1950s and 1960s. Many of the interesting nonlinear theories of that day had algebraic representations involving nonlinear differential equations. These things required the use of numerically intensive algorithms, such as fourth-order Runge-Kutta's and nonlinear least squares estimation routines. Even with today's computers, estimating some nonlinear models can take months at a time. Three decades ago, the situation seemed, and in reality was, hopeless. Thus, researchers had no realistic alternatives if they wanted to conduct empirical investigations. Linear statistical models were the only game in town.

But advances in nonlinear thinking did take place in a setting that later became extremely useful to social scientific analyses. This setting was mathematical population biology. The connection between the social sciences and biology dates back to the early mathematical sociologists, and one particularly general application of biological approaches to modeling social behavior can be found in research by Rashevsky (1954). However, mathematical population biology continued to develop new insights into nonlinear animal and plant behaviors when the social sciences seemed to stop making parallel connections to human society.

In the 1970s, the population biologist, Robert M. May, published two extraordinarily important reports (May 1974, 1976). One of these reports dealt with the behavior of model ecosystems with respect to equilibria conditions, whereas the other was an early discussion of chaos with respect to one-dimensional difference equations. There were perhaps no two other discussions of nonlinearity from population biology that more attracted the attention of social scientists.

The connection between population biology and the social sciences is not as distant as it may appear at first glance. The primary concern in population biology is the interdependencies between various interacting life forms within an ecosystem. The situation in the social sciences is actually quite similar. Instead of interacting life forms, social scientists deal with interacting groups of humans as defined by, say, culture, race, levels of education or status, clan, religion, region, nationality, and partisan affiliation. From a mathematical point of view, these groups interact in ways that are remarkably similar to those that are common to population biology. For example, inner-city gangs act as predators toward their neighborhood populations with regard to the drug trade, and political parties compete as do nations in an arms race.

Finally, major advances in pure and applied mathematics are having a direct influence on nonlinear thinking in the social sciences. Chaos theory is just one example of how nonlinearity is currently making a tremendous impact both in economics as well as other social scientific disciplines. Indeed, it is my sense that the time for a paradigm shift from linear to nonlinear thinking with regard to human societies has finally arrived.

The Linear Hegemony

Typically, when empirical social scientists examine human political behavior, we begin by collecting a set of data. These data can come from surveys of populations taken, say, during election campaigns. Alternatively, the data may be of the aggregate variety, such as county-, state-, or national-level information. These data can be collected over a period of time that is either long or short. Long time spans may extend to centuries, whereas shorter periods can reflect an evening's work for a telephone-based polling company. Once the data are collected, social scientists convert all of the information into numerical form and enter it into a computer for analysis.

The mistake of linearism actually occurs before the data are collected, as I explain below. But this may not yet be apparent at this stage. It is at the next step of social scientific investigation that the mistake of linearism becomes most evident. This is the point when social scientists adopt a mathematical model that becomes the basis for the statistical analyses that follow. This is the model that often reflects, from my perspective, the false linear view of the world.

There are two reasons why social scientists should be so tempted to view human society in linear terms. The first is historical with respect to the computational abilities available with mainframe computers when empirical social science became widespread. But the second reason is that we are now caught in a trap. The way we collect our data reflects a linear view of the world. This view is complemented by the way we conduct our scientific investigations. That is, our methodological procedures closely correspond with a linearized approach to the collection of data.

Let us begin with the historical problem. Between 1960 and 1980, empirical social science leaped forward. In the 1960s, mainframe computers became widely available in all of the major universities. Microcomputers began to appear in significant numbers (even if only one or two in a data lab) in the 1970s. These initial machines, mainframe and micro, were slow. They had small brains (i.e., randomly accessed memory) and limited disk space as well. Social scientists needed to make use of the machines in whatever ways were feasible. The ways that ended up being feasible were those in which analytical solutions were available for the required numbers.

These analytical solutions are formulas. In essence, they are equations that result from algebraic manipulations using calculus. The manipulations begin with a hypothesized function that explains why, say, society is the way it is. This function is called a model. To work with the model, one must have formulas for the parameters that are embedded in the algebra. To get formulas for parameters that were simple enough to use in slow computers, one had to have a relatively simple model. There was also a desire for the models to be sufficiently general such that many scientists could use them without having to derive their own formulas for the parameters.

For one set of formulas to work in many different settings, social scientists had to agree on using one basic model. The choice was almost without competition. The only model that fit all these criteria was a linear one. Formulas could be derived that solved for the parameters of lines (i.e., slopes and intercepts), and the computer code necessary to estimate the numbers from a set of data was not overly complicated. But most importantly, the estimation of the numbers could be done quickly using slow machines.

The second primary reason for the continuation of the linear paradigm in the social sciences, as mentioned above, has to do with the way we collect our data. Linearism is usually thought of as an aspect of our models of the world. But what if we are forced into using linear models because of our data collection practices? That is, what if the models and the data demand each other? How can we break out of the situation? If human behavior is nonlinear, how can we demonstrate this if the data support linear models? Moreover, how could any data support only one type of model?

The problem of obtaining data that are useful for investigating the nonlinearity of human societies is the single most difficult problem facing social scientists today. The problem is twofold, as I explain below. But the consequence of this double-sided dilemma is that we are caught in a trap, the escape from which will demand from social scientists significant levels of ingenuity.

The Trap of Data

We evaluate our theories of the workings of human society based on the data that we have available to us. If our data are structured to favor linear theories, then our results will likely lean in the direction of this bias. There are two problems with most publicly available social scientific data sets that bias our results in a linear direction. The first is that most data sets have only limited temporal measures. Without repeated measures in time, longitudinal nonlinearity is impossible to evaluate. Single cross-sectional surveys (the most commonly available type of data) are useless in this regard. Ideally, to model nonlinear evolutionary processes in human societies, the temporal measures should be both numerous and as closely spaced as possible. Nonlinear studies can be executed with few temporal measures in some situations. But in general, more is better, and sometimes absolutely necessary. Moreover, with few temporal measures it becomes essential that information for each case (i.e., each person interviewed in the survey) appears at each time point, as with a panel survey design.

The second problem with most publicly available data sets is that the information regarding social context is typically extremely limited. For example, survey data sets rarely contain information regarding the neighborhood or county political and social milieu for each respondent. Such information has been added to a few important studies, and these studies act as fundamental windows for the development of contextually based social theories. But it is not the standard practice to make this information available. Contextual information is important from the perspective of nonlinear approaches to social theory since much of the nonlinearity in society is contextually dependent. That is, nonlinearity, both functional and longitudinal, appears more strongly in some social contexts than in others. Theoretical expectations can be established with regard to its manifestation. But without the contextual information, the theories can never be tested.

Thus, data sets useful for nonlinear studies need both repeated temporal measures and contextual information. Moreover, the interaction between these two aspects is crucial. It is not just that some degree of nonlinearity is due to the longitudinal structure of human social and political evolution and a separate degree of nonlinearity is due to context. The two aspects intermix to produce a nonlinear hybrid. It is impossible to evaluate the influence of one aspect without simultaneously controlling for the other.

The proper collection of data for nonlinear studies in the social sciences requires an intimate understanding of the ways in which nonlinearity can be manifest from a behavioral perspective. This addresses our need to identify what we mean by nonlinear.

Two Types of Nonlinearity

Nonlinearity in human societies can occur in two ways.[1] First, the physical structure of the social organizations can have nonlinear characteristics. This is called functional nonlinearity. This can occur, for example, by having people interact with one another. Simply talking and sharing information with others is a nonlinear process. This is one of the processes of contagion and diffusion of information. Simple forms of nonlinearity can be incorporated into the linearized statistical models used commonly today, such as when one variable is multiplied by another. But the nonlinear limits of these models are quickly reached, and more complicated structures simply cannot be written as linear combinations of inputs. Indeed, there are countless ways that societies can be structured (and thus described) in a nonlinear fashion, and this book is an attempt to demonstrate some degree of variety in this regard.

The second way in which nonlinearity can occur is with regard to time, and this is called longitudinal nonlinearity. Things can change in one of two ways. They can change incrementally, in which case a similar increment is added or subtracted to a previous condition in each time period. This could be a situation in which, say, the price of a first-class postage stamp rises by some cents every few years. This kind of incremental change is linear, since the plot of the value of the variable of interest over time in general forms a line. However, nonlinear change is anything else. Change can be based on a percentage of a previous value of a variable, or perhaps a more complicated rule could be used. A simple example of nonlinear change would be the frightening Malthusian (exponential) growth of the earth's human population. In this case, an exponential model is functionally linear, but its path over time follows a curve (i.e., not a line). Models that are functionally nonlinear usually increase the longitudinal nonlinearity of the time path.

The Myth of Independence

The arrival of nonlinear thinking about human life will not occur in the absence of our previous views of the world. Some of these ways of thinking have been fundamental to empirical social scientific thought for many years. Probably the most critical of all of these ideas are the interrelated concepts of endogeneity and exogeneity.

In traditional social scientific thinking, a collection of exogenous inputs causes an endogenous result. Using other language, a dependent variable is found to be a function of a collection of independent variables. These independent variables (the exogenous inputs) have no dependence within them. In a sense, they just exist, and they cause other things to happen. On the other hand, the dependent variable (the endogenous result) is a combination of the influences of the exogenous inputs and some stochastic noise.

From a nonlinear perspective, however, exogeneity may not always be a relevant concept. It is possible to model a relationship in which there is no exogeneity at all. Indeed, the absence of exogenous inputs is actually the norm in nonlinear modeling. What takes the place of exogeneity is system interdependence. System interdependence occurs when one variable is involved in causing change in another, while the other variable is similarly involved in causing change in the first.

Consider the predator-prey model of Lotka and Volterra.

$$dR/dt = aR - bRF \tag{1.1}$$

$$dF/\mathrm{d}t = -cF + eFR \tag{1.2}$$

In this model, R represents the prey (say, rabbits) and F represents the predators (foxes). The term aR expresses exponential growth in the rabbit population, whereas the term $-bRF$ captures the loss in the rabbit population as a function of rabbits being eaten when they unhappily interact with foxes.[2] The term $-cF$ expresses losses in the fox population in the absence of rabbits to eat, and *eFR* represents gains in the fox population resulting from the availability of rabbits. Note that in the above model there are no exogenous inputs. Both the rabbit and fox populations are interdependent in the sense that each causes change in the other. This situation is typical of nonlinear interdependent systems, and this book is filled with examples of such systems.

It would be wrong to interpret my comments here to suggest that nothing can be exogenous. The concepts of exogeneity and endogeneity have been extremely useful to social scientific inquiries, and they will continue to play an important role in our thinking. However, I am arguing for an expanded view of social change, one in which nonlinear interdependence can be seen as an alternative means of expressing this change. The argument is not for a replacement of terms, but for a shift in emphasis, and the reasons for this are not mathematical, but social.

We live within societies that are highly interactive and filled with mechanisms for feedback. Viewing societies in terms of dynamic interdependence simply shifts the primary emphasis in scientific explanations of social phenomena away from static concepts that have correlational associations with dependent variables toward processes of diffusion, contagion, interaction, feedback, and control. It is not that this latter view is correct and the former wrong. Rather, nonlinear interdependence is an expansion of that which we already know. Some social phenomena may be adequately represented as being caused by exogenous forces. Even if the exogenous forces are themselves really endogenous, they may be sufficiently distant from that which is being explained (i.e., the endogenous factor) that the claim of exogeneity causes no great harm. But I argue here that nonlinear interdependence is more common among a large variety of social phenomena than is typically understood in the social sciences generally, and a look in this direction is likely to result in a healthy expansion of social scientific thinking.

Outline of the Book

This book contains four fully developed examples of nonlinear interdependent systems relevant to social science. The substance of each example has been strategically chosen to be quite different from the other examples to enhance the breadth of the current treatment. Thus, each system is addressed separately in its own chapter. However, this book begins with a methodological discussion of a particularly important aspect of nonlinear dynamics that is crucial to the extension of the efforts presented here, given the constraints of data that are currently available to social scientists in general.

Chapter 2 presents a discussion of reasons for using continuous time models of social processes instead of discrete time formulations when working with certain common data situations. The argument is made with respect to data settings with many cases but few time points. Examples of these common data situations would be aggregate units in a country (e.g., counties or districts) for two elections, or individuals surveyed both before and after an election. Whether or not to use a continuous-time model ultimately turns on the question of whether or not the social process being modeled is linear or nonlinear. In general, nonlinear models are much more sensitive to the longitudinal specification of time than linear models. The overall approach to constructing continuous time nonlinear systems has philosophical as well as methodological implications. The continuous time nonlinear systems approach allows greater flexibility in the development of more holistic theories of our societies.

Chapter 3 addresses two discrete literatures in political science. One body of literature is premised on the assumption that voters act rationally with regard to their candidate preferences, implying that voters are individually responsible for their own electoral behavior. Yet another body of literature claims that voters are influenced in their decision-making process by their social and political contexts. In this second case, individuals are seen as being psychologically conditioned by their environment to process information differentially in correspondence with the norms of the society within which the individuals are socially embedded. This latter case embraces a more stimulus-response view of political behavior than the former case, which identifies individual rationality as the primary motivating engine. Some recent studies have attempted to bridge the gap between these two points of view by investigating the limits of individual intellectual autonomy within clearly identified political and social contexts. The current analysis adds to this work by demonstrating that the dual processes of individually rational decision making and the multifaceted interactions of context can produce highly nonlinear dynamics.

Empirically, the investigation in chapter 3 identifies a catastrophe—classically defined in the literature on nonlinear dynamics—as a potential part of these dynamics. The model demonstrates how situations can occur in which individuals can experience very rapid shifts in their feelings toward a particular candidate while simultaneously experiencing relatively small changes in their feelings toward that candidate's party and the opposing candidate. Moreover, the catastrophe component of these dynamics occurs only within a theoretically anticipated type of political milieu.

Chapter 4 investigates mass electoral behavior during the 1964 landslide presidential election in the United States. The aggregate characteristics of landslide elections per se have not been examined thoroughly in the extant literature on voting. Here, a nonlinear model of partisan competition is developed and evaluated with respect to a complete collection of county-level electoral data. The model is a system of two interdependent differential equations characterizing aggregate rapid and large-scale partisan change. It is found that the 1964 landslide election involved a highly complex and contextually conditioned set of aggregate voting behaviors. The masses were guided in their partisan choices by a variety of nonlinear social processes. Moreover, in the deep southern states, the process of partisan competition was not completed by the time the election occurred.

Evidence is offered in chapter 4 that suggests that the electorate in the Deep South did not vote in a state of regional equilibrium. The opposite is true of aggregate voting in areas outside of the Deep South. These findings have implications with regard to the meaning of elections during periods of rapid partisan change. Finally, survey data are examined for that election to complement the aggregate findings with a psychological interpretation of the 1964 campaign. It is found that voters discerned real ideological differences between Johnson and Goldwater, and that these differences acted as the trigger to initiate the mass partisan movements characteristic of that landslide.

The analysis in chapter 5 introduces a formal nonlinear systems model of context and voting during the later elections of the Weimar Republic. The interdependencies involve three variables: (1) the level of aggregate vote shifting between the parties (i.e., an operationalization of deinstitutionalization), (2) the level of partisan fragmentation of the electorate (operationalized using Rae's fragmentation index), and (3) the level of support for the Nazis. The analysis reveals a high level of nonlinear complexity between the variables. Moreover, the discovery is made of nonlinear catastrophe superstructures within which variable change is imbedded. This suggests that electoral systems may experience localized evolution, with observed changes in the variables being orchestrated by large-scale—and often unnoticed—highly complex nonlinear structures. These large-scale structures are extremely important since their size and shape can change over time, resulting in the sudden insertion of nonlinear equilibria surfaces within the realistic ranges of actual variable values. The Weimar Republic experienced such a situation, and its fall can be characterized in large part as a consequence of a rapid distortion of a previous, and fragile, balance within the electoral system.

Chapter 6 presents an analysis that investigates longitudinal change in the environment as a function of oscillating partisan control of the White House. It is assumed that one political party will tend to favor help for the environment despite some economic costs, whereas the other party will generally favor economic growth over environmental concerns. These policy changes affect the environment interactively with (1) public concern for environmental problems and (2) the economic costs relating to environmental repair. This interaction with policy changes causes a disruption in the continuously evolving balance between the social factors that damage the environment and the environment's own ability to recover. The disruptive potential to the environment is considerably ameliorated with a reduction in the electoral cycling.

The final chapter of this book presents ideas that could help guide the development of a more general theoretical perspective of nonlinear political and social evolution. This chapter is the most speculative in the current context. However, some effort is needed to add synthesis to a volume that covers as much substantive and methodological territory as this one. It is important to point out that the substantive examples that are explored here do tie together, and they were not chosen in an ad hoc fashion. One of the primary purposes of this book is to demonstrate how much can be gained substantively from a nonlinear dynamical systems perspective of the interdependencies of human social life. One extended example simply would not accomplish this task. The broader point is that we live in a nonlinear universe, and humans are not the linear exceptions to the nonlinear norm. What is at stake is a perspective of human existence, and a broad array of substantive topics within the field of political science is needed to make this case.

This does not say that the current volume is meant to be a definitive treatment of nonlinearity and social phenomena. Rather, this book serves as an intellectual marker that may stimulate further thought as to how we humans should continue to view ourselves and the evolution of our societies. In a very real sense, this book is meant not as an end but as an extension of a perspective of human development that has already started, and is evidenced in the small but growing literature on the interdependent complexities of our societies.

Social phenomena are like serpents. Not much is linear about them, particularly with regard to their movements through time. They follow not a direct path, but wind among the vagaries of the development of the broader human culture. This broader culture also is not fixed. It is like sand. The marks that are made on it by social phenomena quickly blend into the historical evolution of this culture, mixing with the previous marks of all other phenomena, until it is difficult to say that any one mark is distinct from all of the others. We live in a human universe of movement and change. That which seems fixed may be only the ghost of a contrived static perspective that ignores the continuing evolution of human interdependencies. Indeed, the fundamental proposition of this book is that most of our experiences are those of change within a background of further change; our lives developing like the movements of serpents, serpents in the sand.

CHAPTER 2

The Structure of Nonlinear Time

The social sciences are now poised on the threshold of a new era in mathematical modeling and data analysis using continuous time nonlinear dynamical systems. Some such models, such as chaos and catastrophe models, have been recently popularized in the mathematical and scientific literatures. Moreover, there have been some fascinating attempts to develop exploratory methods of data analysis to identify situations in which such models may be useful in social scientific settings (Richards 1992, McBurnett 1995a, 1995b). Progress has also been made to develop and estimate continuous time nonlinear dynamical systems (Brown 1991). To see this as a threshold to a widespread new era in modeling, it is necessary to understand the data constraints that have made the use of continuous time nonlinear systems so infeasible for social scientists in the past, and then to see why these constraints no longer exist.

Social scientists typically encounter data situations of very short time series. These series are often so short—two or three time points—that they do not always think of them as time series. They call them panel studies, or preelection and postelection surveys, or aggregate data for two or more elections. But these really are time series, and the technology of nonlinear dynamical systems is now sufficiently advanced that we can begin to model these data in ways that we might otherwise have thought were limited to the natural sciences, or at least to relatively rare data situations with many observations taken over an extended period of time. The key to developing nonlinear system models for these data for extremely short series is to utilize the information that is contained in variations among many cases.

Multiple cases—e.g., respondents in a survey, counties or districts in a country—over two time points are perhaps the most common data situation encountered by social scientists. On the other hand, the literature of dynamical systems is generally oriented around explaining the longitudinal characteristics of a single item. That item can be a pendulum, a rocket, a nation, or even a planet. With that item are associated a number of variables. A dynamical system then characterizes longitudinal change in these variables for that one item. For example, a dynamical systems model for a pendulum can have angular velocity, a phase drive term, and other variables. A rocket can have acceleration, fuel, and air resistance. A nation can have defense spending and social spending. A planet can have pollution as well as efforts to clean up pollution. For all of these examples, we have one case (i.e., item) with a few variables. In the past, typical time series problems in the social sciences have fallen into this category as well. The critical data problem then becomes obtaining a sufficient number of observations across time (and over a sufficiently long period of time) in order to specify the model and its long-term dynamical structure.

Thus, on the surface, it seems that a data situation of many cases across two (or a few) time points is incompatibly different from a situation with one case with a long observable history. The trouble seems to stem from the need to have a sufficiently long data record to enable the determination of the fixed dynamical structure. Yet the same dynamical structures, at least in theory, should operate with many cases just as they do with one. The fact that the number of observations over time are few should not deter us from perceiving that a similar dynamical structure may lay hidden between those few time points as well.

In the social sciences, the usual approach to such data situations is to employ a statistical regression model. Such a model requires a dependent variable that can be written either as a steady state or as a difference between two steady states. Thus, if one has aggregate electoral data for 3,000 counties in the United States for two elections, one can take the difference in a party's vote between those two elections as a dependent variable. Two functional forms are the most common, as are specified in equations 2.1 and 2.2.

$$\Delta y_t = F[y_t, x1, x2, \ldots, x(n)] \tag{2.1}$$

$$y_{t+1} = G[y_t, x1, x2, \ldots, x(n)] \tag{2.2}$$

Since $\Delta y_t = y_{t+1} - y_t$, these functional forms are essentially identical with the exception of how the dependent variable is written. For example, if equations 2.1 and 2.2 are linear models, the slopes for all of the $x(i)$ variables would be identical, and the slope for y_t in equation 2.2 would be one plus the slope of y_t in equation 2.1. Of course, the R^2 for each model would differ since this statistic is sensitive to the longitudinal structure of the dependent variables due to the relative magnitudes of change in y_t versus that for cross-sectional variation in y_{t+1}. Similar arguments, varying only in the level of algebraic complexity, could be made for functionally nonlinear models as well. Since equations 2.1 and 2.2 are essentially identical, for the purposes of this discussion, attention will be focused on measuring the dependent variable as a difference (i.e., eq. 2.1), leaving it up to the reader to make any wanted connections with models in the form of equation 2.2.

There are clear logistical advantages of using Δy_t as a dependent variable. The difference is easy to calculate using any of the commonly used statistical software packages. Moreover, one can often use some form of multiple regression to make the required estimations. These reasons are not sufficiently valuable, however, to risk losing the ability to discern the accuracy of one's model. Yet this is precisely what can happen if one is using a discrete measurement of a *nonlinear* continuous social process. To show how this can occur, it is important first to discuss why the use of a discrete measurement does not similarly distort the model estimations of a *linear* continuous process.

Continuous Time with a Linear Model

Equation 2.1 could easily be rewritten as a continuous time model. In this case, the dependent variable would no longer be Δy_t, but rather the derivative expressing change in y over time, dy/dt. For the moment, we can consider the case of a functionally linear model using continuous time, as given in equation 2.3. Moreover, we will compare equation 2.3 with its discrete counterpart, equation 2.4.

$$dy/dt = ay + b \tag{2.3}$$

$$\Delta y_t = fy_t + g \tag{2.4}$$

Both equations 2.3 and 2.4 are functionally linear and identical in their algebraic structure. The only difference between the two equations is with regard to the measurement of time. However, to use equation 2.3 in a practical sense, it is necessary to obtain a predicted value of y for a given point of time. Let us say that we have two observations of y. The first observation we can call its initial value, whereas the second observation we can identify as the endpoint. With a continuous time model, we need to use the model to predict the endpoint given the initial condition.

It is possible to solve for y in equation 2.3, thereby obtaining an equation for y as a steady state that could be used as a dependent variable in a regression model. However, this is not always possible with interesting nonlinear models. The situation becomes much more intractable mathematically with nonlinear systems of equations. Thus, we want to solve for y in the most general way possible, a way that will work for nonlinear models and systems as well. One way, and perhaps the easiest and most useful method, is to use a fourth-order Runge-Kutta technique of definite integration. This technique, while not often used in the social sciences, is the standard workhorse for numerical analyses in the natural sciences. A helpful introduction to this technique can be found in Hamming (1973), while extensive use of this and related numerical methods in the social sciences can be found in Brown (1991). It is important to note, however, that none of the results in this chapter are dependent on the use of a Runge-Kutta algorithm. Other algorithms for use in definite integration are available and would yield similar results. Runge-Kuttas are the most widely used of such methods, however.

It is important for readers to keep in mind throughout this discussion that we are interested in modeling change between two (or at most a few) time points. But Runge-Kutta methods allow us to use a continuous time model to trace a trajectory from one observation to another as if there were a smooth flow of observations between the initial observation and the endpoint. This is a highly desired aspect of continuous time models since, if we can do this with some degree of confidence, we can recreate some unobserved historical change between the time points. Moreover, these trajectories can be highly nonlinear regardless of the linear or nonlinear nature of the functional form of the model. This type of nonlinearity is called longitudinal nonlinearity and is explained more fully below (see also Brown 1991, 29–36). The ability to recreate longitudinally nonlinear trajectories between observed time points is one of the few advantages of continuous time linear models over discrete time linear models. This can be seen with a direct comparison between equations 2.3 and 2.4.

Table 2.1 presents a comparison between equations 2.3 and 2.4 with data constructed using heuristically chosen parameters for the continuous time model (eq. 2.3). The parameters for the difference equation (eq. 2.4) are estimated from the initial and endpoint data presented in the table. Thus, the initial values and the differential endpoints were used to calculate Δy_t. This difference was then used in a regression model to obtain values for the parameters f and g. The differenced endpoints in table 2.1 were obtained from the difference equation after the estimations were completed.

In table 2.1, note that the discrete linear model can exactly reproduce the endpoint data that was created by the differential equation model. Moreover, the fit of the difference equation is unity, and the parameters f and g have the correct signs as compared with their parallel continuous time counterparts. Also, note that the relative magnitude of parameters a and b is exactly the same as the relative magnitude of parameters f and g (one magnitude is double that of the other). Thus, if someone was to draw substantive conclusions from these models, there would be little lost in using the discrete time version.

What would be lost is the nonlinearity of history between the two time points. This can be easily seen in figure 2.1. In figure 2.1, the continuous time trajectories for equation 2.3 are computed using the parameter values presented in table 2.1 and a fourth-order Runge-Kutta algorithm with 10 iterations and a step size of 0.1. Note that the movement between the initial values of y and the endpoint values of y does not follow a straight line. In some

TABLE 2.1. Comparison between Two *Linear* Models with the Same Algebraic Form but with Different Time Structures, One Continuous and One Discrete

Differential Equation	Difference Equation (estimated)
$dy/dt = ay + b$	$\Delta y_t = fy_t + g$
$a = -2$	$f = -0.864$
$b = 1$	$g = 0.432$
$y^* = -b/a = 0.5$	$y^* = -g/f = 0.5$
	$R^2 = 1$

Ten Paired Initial and Endpoint Observations.

Initial	Differential Endpoints	Differenced Endpoints
0.1	0.446	0.446
0.2	0.459	0.459
0.3	0.473	0.473
0.4	0.487	0.486
0.5	0.500	0.500
0.6	0.514	0.514
0.7	0.527	0.527
0.8	0.541	0.541
0.9	0.554	0.554
1.0	0.568	0.568

Note: Values for y^* in tables 2.1 and 2.2 are equilibria values.

research, this historical re-creation may be important, in which case, the continuous time model has the distinct advantage over the discrete time model. However, if we are only interested in explaining the total change in y, then there is no harm in using the discrete time model, *assuming that the social process being examined is functionally linear.*

The reason for this conformity between the continuous and discrete linear models can be extracted from the solution forms for equations 2.3 and 2.4. The solution forms for each of these models are given below as equations 2.5 and 2.6, where equation 2.5 represents the solution form for the continuous time model and equation 2.6 represents the solution form for the discrete time model (see also Goldberg 1958). In each solution form, y^* represents the equilibrium value for the model. This value is the same for both models since we wish to choose parameter values that will predict endpoints that will exactly match at integer points in time. The initial values of y are represented by y_0 in both equations.

$$y_t = y^* + (y_0 - y^*)e^{at}, \qquad a \neq 1 \tag{2.5}$$

$$y_n = y^* + (y_0 - y^*)(f + 1)^n, \qquad f + 1 \neq 1 \tag{2.6}$$

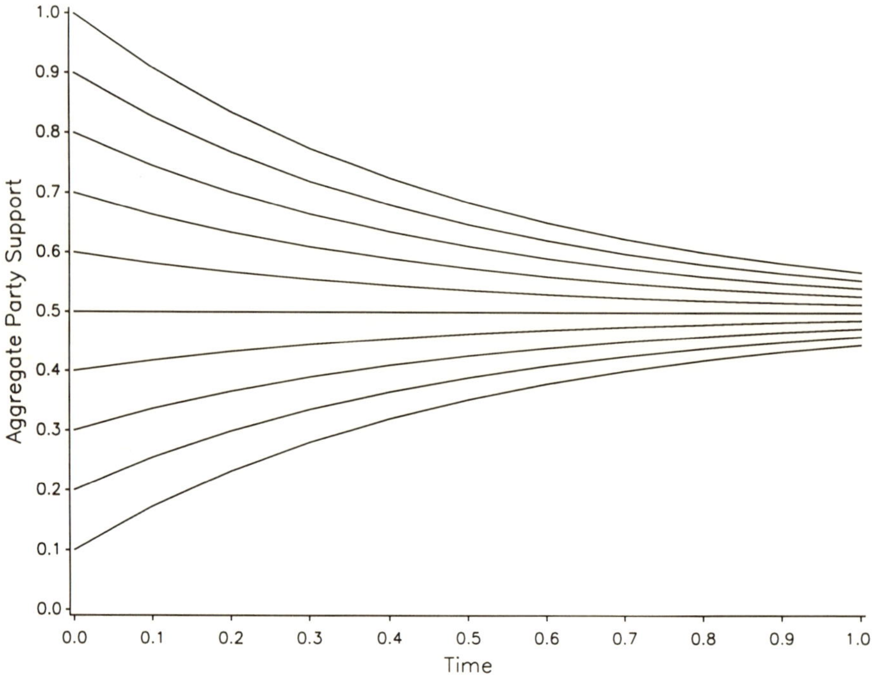

Fig. 2.1. Sample trajectories for a functionally linear continuous time model

The error between the continuous and the discrete models is the difference between these two solution forms. Subtracting equation 2.6 from equation 2.5 and then simplifying (noting that $t = 1$ at the endpoint) yields the expression $(y_0 - y^*)[e^a - (f + 1)]$. This expression equals zero as long as $e^a = f + 1$. Moreover, this condition is entirely independent of the initial conditions of y. Since e^a is a constant, regression has no difficulty determining a value for the parameter f that exactly satisfies this condition, thereby maximizing the fit of the model to unity. This lucky algebra is, in general, absent from the case of the functionally nonlinear model as is explained below.

Continuous Time with a Nonlinear Model

The incompatibility between continuous and discrete time nonlinear models is best explained through an example. Perhaps the simplest and most well known nonlinear model is the logistic equation. Continuous and discrete time representations of the logistic equation are given here as equations 2.7 and 2.8, respectively.

$$dy/dt = y(a - by) \tag{2.7}$$

$$\Delta y_t = y(f - gy) \tag{2.8}$$

Table 2.2 contains a comparison between equations 2.7 and 2.8 using heuristically chosen parameter values for the continuous time model. The parameter values for the discrete time model (i.e., equation 2.8) are estimated

TABLE 2.2. Comparison between Two *Nonlinear* (logistic) Models with the Same Algebraic Form but with Different Time Structures, One Continuous and One Discrete. Also, the Parameter Estimates for a Logistic Model with an Intercept

Differential Equation	Difference Equation (estimated)
$dy/dt = y(a - by)$	$\Delta y_t = y(f - gy)$
$a = 3$	$f = 0.839$
$b = 6$	$g = 1.428$
$y^* = a/b = 0.5$	$y^* = f/g = 0.588$
	$R^2 = 0.82$

Ten Paired Initial and Endpoint Observations.

Initial	Differential Endpoints	Differenced Endpoints
0.1	0.436	0.170
0.2	0.474	0.311
0.3	0.488	0.423
0.4	0.495	0.507
0.5	0.500	0.563
0.6	0.503	0.589
0.7	0.505	0.588
0.8	0.507	0.557
0.9	0.508	0.499
1.0	0.509	0.411

Difference Equation with Intercept (estimated)
$\Delta y_t = y(f - gy) + k$
$f = -0.781$
$g = 0.142$
$k = 0.427$
$y^* = 0.501$
$R^2 = 0.999$

with regression using as data the initial values of y and the endpoints of y that are obtained from the differential equation after extending the differential equation with a fourth-order Runge-Kutta 10 iterations with a step size of 0.1, exactly as was done with the linear models as presented in table 2.1. For purposes of comparison, table 2.2 also presents estimated parameter values for a related discrete time model, a logistic model with an intercept, which is explained fully below following the comparison of equations 2.7 and 2.8.

From table 2.2 it is clear that equations 2.7 and 2.8 do not produce the perfect match that was seen in table 2.1 with regard to the linear models. With the nonlinear models, the predicted endpoints are not nearly the same. Moreover, while the parameter values all have the correct sign, their relative magnitudes do not exactly correspond across time equations. That is, the value of parameter b is twice that of the value of parameter a, but the value of parameter g is not twice that of the value of parameter f. The equilibria values for both models are not the same. Finally, the R^2 for the discrete model is only 0.82. Clearly, something is not working the way it worked with the linear models.

The situation is clarified by an examination of figure 2.2. Figure 2.2 presents ten trajectories of the continuous time logistic model over the time period from 0 to 1. Unlike the trajectories of the linear model presented in figure 2.1, the approaches to the equilibrium value of 0.5 using the logistic model are not proportionally symmetric with regard to whether the initial values are above or below the equilibrium. As a general rule, the more nonlinear the model, the more intricate will be the longitudinal dance of the trajectories, and the less proportional symmetry there will be with regard to the approach to equilibria values.

Intuitively, it should be clear that a model that is extended only one iteration between two widely spaced time points—as is done with the difference equation model—generally cannot arrive at the same endpoints as a model that is given the flexibility of continuous nonlinear movement in time. With the discrete time model, the rates of change are dictated by the initial values of y, whereas with the continuous time model the rates of change vary continuously—and nonlinearly—as the values of y evolve. A fixed rate of change cannot produce the same result as a variable rate of change over the same time period given the lack of topographical symmetry that is inherent in the behavior of nonlinear trajectories.

The structure of the error between the continuous and discrete time models can be seen clearly in figure 2.3. In figure 2.3 the final values of y (i.e., the endpoints) are mapped onto the initial values of y for both equations 2.7 and 2.8. Note in this figure that the differential equation endpoints rapidly converge to the equilibrium value of 0.5 as the initial condition increases. However, the difference equation endpoints reveal a parabolic distribution

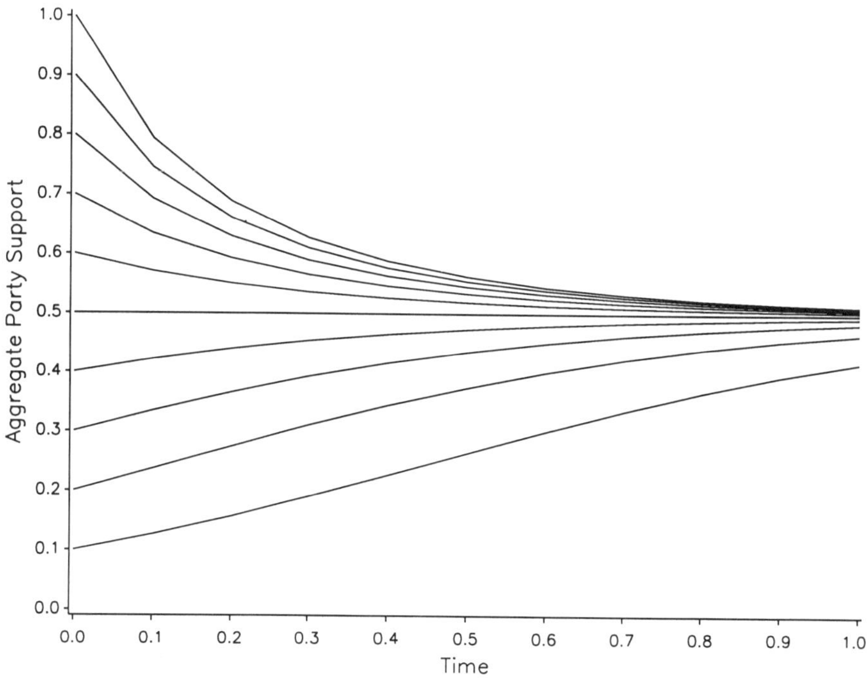

Fig. 2.2. Sample trajectories for a nonlinear (logistic) continuous time model

that results from the quadratic nature of the logistic equation in combination with the inflexibility of using this equation through only one iteration. The difference equation model does the best it can do given its limitations, but the best it can do cannot overcome the problem imposed by its rigidity with regard to time.

It is essential for readers to note that the current discussion refers only to situations in which there are only two time points, and in which there is a choice of using either a difference equation or a differential equation to model change between those two time points. The argument here has nothing at all to do with the use of a particular numerical algorithm such as a fourth-order Runge-Kutta. The Runge-Kutta is merely used here as a convenience to trace out a nonlinear trajectory between two points. In the above examples, it is used to create a longitudinally nonlinear data set given known parameter values for a continuous time function. These data could have come from any other source (including angelic) with no loss to the argument. I am not saying that Runge-Kuttas are intrinsically different than difference equations. Runge-

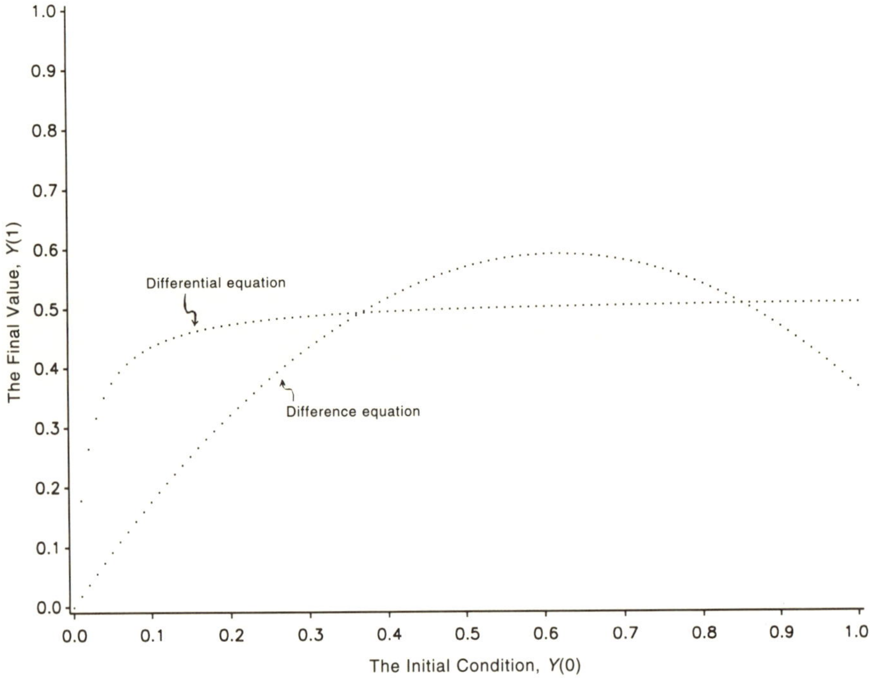

Fig. 2.3. Selected end points for continuous and one iteration discrete time logistic models

Kutta algorithms are, indeed, difference equations with a variable step size, whereas traditional difference equations work only on the integers of time. But precisely because of the variable step size, we can use these techniques to trace out the longitudinally nonlinear trajectories of continuous time models. This is a convenience to my argument, not a central point.

The points being made here, however, have everything to do with the longitudinal specification of the model. Differencing between only two time points loses all nonlinear longitudinal variation. If the social process is a short-term nonlinear continuous one, using a nonlinear model over a widely spaced interval, as is done by differencing a dependent variable and employing regression technology to get the estimates, can lead to a serious misunderstanding of the dynamical process.

Thus, when comparing the differential versus differenced results presented in table 2.2 and figure 2.3, it is perhaps heuristically useful to ignore the particular means by which the differential endpoints are calculated so as not to be confused into thinking that the results are a product of differences between the two algorithms. Consider the continuous time data merely as accurate measurements of a trajectory originating from a nonlinear process about which a totally accurate specification is known. Then make the comparison with these data and those created using the estimated difference equation. The point is that the real endpoints (coming from the continous time process) could not be re-created by the nonlinear but differenced model. Yet in the social sciences, that is precisely what so often occurs. A nonlinear continuous time process is measured at two widely spaced intervals, the variable measurements are differenced, and a model is evaluated!

The basic problem between the continuous and discrete time nonlinear models can now be summarized. With regard to the continuous time nonlinear models, there are two types of nonlinearity at work. The first is functional nonlinearity, which is due to the algebraic structure of the model (e.g., in the logistic case, the multiplication of y times y). But the second is longitudinal nonlinearity. Again, longitudinal nonlinearity refers to the nonlinear movements of model trajectories over time, and even linear dynamic models can behave nonlinearly in this regard, as was seen in figure 2.1 (see also Brown 1991, 29–32). With the continuous time nonlinear model, both forms of nonlinearity interact to produce the trajectories. But the discrete time models are given only one iteration (since there are only two time points in the research setting under discussion), and thus their abilities to produce longitudinal nonlinearity are nonexistent. Indeed, all discrete time models can produce only a straight line between any two time points. Thus, we have a classic case of an underdetermined problem. One form of nonlinearity (functional nonlinearity) with a discrete time model is being used to reproduce what is created by a continuous time process that interactively employs two types of nonlinearity (functional and longitudinal). If the social process that is being examined is of a continuous time nature—and it can be argued that most social processes are—then the discrete time model over two time points will likely suggest too low a fit and seriously incorrect parameter estimates, *despite the fact that the algebraic functional form of the model may be absolutely correct.*[1]

However, the problem may be worse than that. It is common practice in most social scientific statistical analyses to employ an intercept term in the model. Typically, the inclusion of an intercept is thought of as a relatively minor matter of model specification. Here it is used as an example of how a small variation in model specification can produce very misleading results when a discrete time nonlinear model is used to evaluate a continuous time social process.

At the bottom of table 2.2, the parameter estimates are presented for a logistic difference equation that includes an intercept term. The following observations of the results are particularly relevant here. The magnitude of the intercept term is quite large. The sign of the estimate for parameter f is incorrect. But the fit of the model is near unity and the equilibrium value for the model is nearly 0.5!

These results are further developed with an examination of figure 2.4. Figure 2.4 presents various trajectories (corresponding to different initial conditions) for the logistic difference equation with an intercept. Note that the trajectories in figure 2.4 rapidly converge in one iteration to values that are similarly distributed around the equilibrium value as compared with the endpoints of the continuous time model used to create the trajectories presented in figure 2.2. Moreover, the trajectories continue to converge to the equilibrium value following the first iteration of difference equation.

From a modeling point of view, the situation described here is quite unfortunate. If one is using a nonlinear discrete time model to evaluate a nonlinear continuous time process, a minor specification error—such as the inclusion of an intercept term when in fact the social process does not reflect one—can lead to results that seem very strong from a statistical perspective.

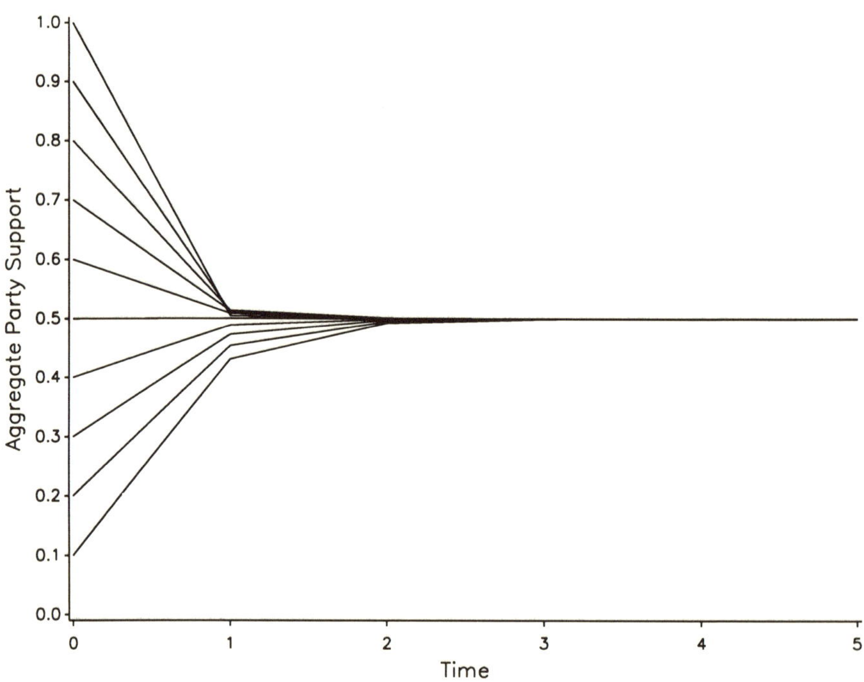

Fig. 2.4. Sample trajectories for the discrete time logistic model over five iterations

Yet the results could be misleading at best, and totally incorrect at worst. One problem, functional misspecification, can interact with another problem, longitudinal misspecification, to confound the statistical evidence of both mistakes!

The only solution to this dilemma is to return to the continuous time nonlinear model when modeling a social process that evolves in a more or less continuous fashion. The question then becomes one of how to engage the mechanics of continuous time estimation given the lack of a discrete time alternative. An overview of this is included in the appendix to this discussion.

In general, this matter of longitudinal specification focuses on the type of social process being modeled. If a researcher is certain that the social process is dynamically linear, then no great error is made in measuring change in the dependent variable as a discrete difference and then using some form of multiple regression to make the parameter estimates. *But one must make this assumption of linearity in order to proceed with confidence*. On the other hand, it is perhaps more typical not to know this for certain in advance of engaging in an analysis of the data. Therefore, one may be on safer ground in not assuming linearity, and accept with this the freedom of working with nonlinear functional forms, as well as the possibility of uncovering surprises—and perhaps beauty—in diverse portrayals of longitudinally nonlinear social change.

Nonlinear Dynamical Systems

Leaving the familiarity of differencing and regression gives an additional benefit. There is no need to work only with single-equation continuous time models any more than there is a need to be restricted to linear approximations of nonlinear processes. The numerical methods used for single-equation models are identical to those used for systems of equations. Thus, we can talk about the nonlinear evolution of social systems as easily as we can talk about longitudinal change in one variable.

On one level, working with nonlinear systems is easier than working with many linear regression models. One need never worry about transforming complicated functional forms into algebraic versions that can fit within an existing model that comes "canned" in statistical software. Nearly all numerical methods that are used for continuous time systems require only that the model be written as a derivative. Indefinite integration is not necessary. Transformations are almost never necessary. Most importantly, distortions of the social theory due to the algebraic requirements of statistical models are totally unnecessary. Thus, one is naturally encouraged to be intellectually creative when using nonlinear continuous time systems.

But with this encouragement comes a warning. Since a nonlinear view of the world appears much different from a linear view from an algebraic perspective, familiarity with linear models must not bias one's judgment of nonlinear algebraic structures. The problem is especially acute with regard to the way in which the generality of nonlinear systems is perceived in situations of algebraic asymmetry across equations. Such systems may at first appear ad hoc, when in fact they can be theoretically rich and quite general.[2]

Symmetric nonlinear systems are those in which each equation in the system has a parallel algebraic structure. This is often the situation used to model competition, as between political parties. For example, it may be assumed that both parties are competing for the same resource (i.e., voters) in the same way. Thus, the algebra describing change in each party's support would be identical in structure. In practical terms, we can build a model of competition that has a logistic structure on the rates of growth for each party combined with losses for each party that are due to proportional interactions with the other party. These characteristics are similar to many interspecies competition models that are described in the extensive literature on population biology (see May 1974). Thus, we can describe change in the support for each party as in equations 2.9 and 2.10.

$$dx/dt = ax(L_x - x) - bxy \tag{2.9}$$

$$dy/dt = cy(L_y - y) - exy \tag{2.10}$$

In equations 2.9 and 2.10, the limits of growth for parties x and y respectively are L_x and L_y. The parameters a and c are the logistic growth rate parameters, whereas the parameters b and e define loss for each party due to proportional competitive interaction with the other party. Note that each equation has an exactly parallel algebraic structure. Sometimes a system such as this appears more general than an asymmetric system (described below) since the symmetric system does not unequally restrict population interactions and exchanges between any group and any other group (e.g., party).

However, it is likely that many social systems will not display algebraic parallelism across equations. Such systems are asymmetric in the sense that change in one of the variables is not similarly structured to change in one or more of the other variables. The familiar Lotka-Volterra predator-prey equations are one such system, as presented previously as equations 1.1 and 1.2.

The system represented by equations 1.1 and 1.2 clearly has an asymmetric algebraic structure across equations. On the surface, the predator-prey system may seem more ad hoc due to this asymmetry. But this is not the case. Algebraic richness tied to symmetrical or asymmetrical model properties has no implication for the generality or the theoretical value of the specific model. Indeed, the necessity of the asymmetry is dictated by the physical structure of the natural phenomenon. Moreover, the approach to generality is best viewed from the perspective of the model's overall behavior. This overall dynamical behavior is likely to be categorizable from within a broad range of behavioral geometries (see Abraham and Shaw 1992). The contribution of the symmetry characteristics of the system's algebra to its overall behavior is thereby generalizable from within this scheme of categorization (see, especially, Hirsch and Smale 1974, 258–75).

The "bottom line" is that nonlinear systems can be either symmetric or asymmetric with regard to algebraic parallelism and still be quite general from the perspective of systemic behavior. The value of both functional and longitudinal nonlinearity rests in its ability to describe more accurately the complexity of the relationships between variables.

It is probably not just good luck that this complexity seems more often than not to have aesthetically pleasing qualities as well, especially when the behaviors of the systems are portrayed graphically (a point often made in the literature on fractals). More likely, this is because the natural beauty of the phenomenal world is a consequence of the very same structural characteristics that nonlinear specifications attempt to identify.

Taken one step further, this suggests that it may occasionally be useful to view a social scientific theory not only with respect to the accuracy of its portrayal of relationships between variables, but also with respect to the magnitude of the beauty that is captured in that portrayal. Indeed, it is interesting that such beauty in nonlinear behavioral relationships between and among people may only be discernable through mathematical—and thus graphical—means.[3] It may at first seem odd for social scientists to don the garb of the artist. But beauty does exist in nature, even though we have no statistical test for it and its evaluation is ultimately subjective. The point here is that nonlinearity offers a way to identify some of the important realities of human relationships, and we should not be blind to the possibility that a portrayal of beauty—however defined—may be captured in the nonlinear specification in ways that a linear approximation could never match.

Some Implications

Perhaps the greatest implication of the use of nonlinear systems in the social sciences is that the methodological divisions between the natural and social sciences are greatly diminished. After reviewing virtually all of the commonly utilized statistical approaches to modeling and data analysis in the social sciences and comparing them to the approaches common to, say, population biology, physics, astronomy, and chemistry, one might naturally wonder why human beings are the only creations in the known universe that behave linearly. In reality, human beings probably behave no more linearly than the orbit of the planets, and the use of nonlinear systems to model social change allows us to examine this proposition directly.

I end this discussion on an evangelical note. If one takes the implications of the need to pursue continuous time nonlinear systems seriously, there exists a tremendous opening to new intellectual horizons for social scientists. This goes beyond returning to previous studies and reworking the methods in the hope of changing the conclusions. Rather, the techniques themselves encourage social scientists to search for fixed dynamical structures in our societies. It even leaves the door open to the discovery of principles—analogous to "laws"—of human activity, a goal that has too often eluded social scientists.

More specifically, the use of nonlinear systems opens the door to thinking about humans as social beings rather than merely (or at least predominantly) as isolated, atomized, rationalizing, or psychologically driven individuals. For example, when the social sciences embraced the benefits of social surveys with national probability samples, it was not without cost. Surveys typically extract the respondents from their social environment and ask them questions about their ideas and feelings. It is no wonder why psychological and rational theories are so common in the social sciences, given the nature of such methods that are typically used to examine social questions. Questioning individuals about things in their minds leads to theories driven by individual minds.

Future research in the social sciences will almost certainly have to reconnect our understanding of individuals with a holistic understanding of our societies. Just as studying one wave in the ocean is incomplete (since all waves affect all other waves), studying individuals extracted from the ongoing and continuous time feedbacks of social existence is similarly incomplete.

If this hubristic argument is accepted, the question of social determinism will likely be raised eventually. Underlying this perspective on methodology rests a view of human existence as fundamentally interconnected on an evolutionary scale both internally (i.e., among and within the community of humans) as well as externally, as with our environment and planet. The future may find human activities less close to an individually defined probabilistic view of choices based on rules of decision making and a collection of certain attitudes, and closer to a consequence of nonlinear and highly interactive feedbacks of stimuli collectively defined. It is not that rationality, psychology, and other concepts of the individual will no longer play a part in the social theories of the future. Humans do calculate and have attitudes. It is just that future paradigms may more clearly recognize the social constraints that restrict in real terms the range of individual activity, intellectual or otherwise. This implies that we may be able to construct broad generalizations of the structure and evolution of our present and past communities on a scale that was not possible before.

Thus, the threshold that I mentioned at the beginning of this discussion refers not just to the use of some new mathematical tricks. It refers to the way we conceive of human existence. I suggest that the new methods will allow a greater flexibility in our thinking about what it means to be human. In this sense, social scientists may be on the verge of a future as exciting as that referenced by Robert May (1974) for population biologists when he outlined his now classic view of ecological structure and niche overlap. This is not a prediction, but a speculation, however brash. Yet I suspect it may be true.

APPENDIX TO CHAPTER 2

We return now to the initial statement of the problem. Social scientists typically encounter data situations in which there are many cases but only two (or a few) time points. This discussion points to the possibility of using nonlinear continuous time models to bridge the gap between the two time points. What one needs (besides a nonlinear model) are parameter estimates.

Estimation details of nonlinear models of the type discussed here (including the script of a workable computer program that is included in the appendix of this volume) have been discussed extensively elsewhere (see Brown 1991, 203–15; 1995). Nonetheless, it is useful to emphasize the general outline of the procedure here.

Briefly, to estimate continuous time nonlinear models with many cases but few time points, one needs to use a Runge-Kutta technique to create a trajectory from the initial values of the data (i.e., for each case) to the final values. Since there are many cases, this requires the use of a matrix programming language so that all cases can be handled at the same time. Usually one can use a step size of 0.1 and 10 iterations of the Runge-Kutta algorithm to accomplish this. If more than two time points are involved, then the number of iterations between each pair of time points should be proportional to the relative length of each time segment. For example, if the actual time between the first two time points is one year whereas the actual length of the subsequent time segment is six months, then one may choose to use 10 iterations between the first two sets of points and to continue with 5 iterations between the second two.

The initial values of the parameters for the model are guessed using a randomization procedure. The fit of the model to the data is then evaluated. This is usually done with respect to explaining change in the dependent variable rather than cross-sectional variation in the final value of the dependent variable. At this point it is necessary to change the values of the initially guessed parameter values in order to maximize the fit of the model to the data.

The key point is that we are able to recreate the most likely nonlinear trajectory (for which there are no actual data observations) that connects the data values across time. This is possible because of the pairwise variations in the placement of initial data values as well as the endpoint values.

CHAPTER 3

Individual Voter Rationality and the Influence of Context: The Relationship Is a Catastrophe

The interactions between individuals and their political environment are activities that are exceptionally rich with nonlinear potential. Citizens seek to determine appropriate behaviors for themselves when they cast their ballots. They consider their options and evaluate their preferences. They look at, and listen to, the candidates for public office. They also talk to their co-workers and neighbors to solicit additional perspectives and to offer their own. They interact with other politically active individuals in stores and on the streets when they engage in conversations, react to political television advertisements, witness the wearing of political buttons, note the erecting of yard signs, and drive behind cars with political bumper stickers. In countless ways, the environment in which we all live supplies us with political information that either directly or indirectly informs us of the bias of that environment. That bias partially conditions us, as individuals, to either accept or reject political information that may or may not conform to our previously held views. Or, perhaps, we may simply note that the bias exists, letting it slowly shape the way we view our world, not necessarily in a fashion that is entirely obvious to us.

All this implies the existence of feedback. We have opinions, we give opinions, we listen to the opinions of others—however they may be delivered. We influence others as much as we are influenced by others. Even our influence on others reinforces our own behavior, and the process almost never ends. Indeed, we both give and receive information on a near-continual basis. Moreover, the feedback need not be limited to close personal contacts, since the broader social environment also flavors our daily life.

This chapter presents research in which an extreme form of nonlinearity is found to characterize the dynamic interactions between voters and their more general political milieu. Elsewhere in this book, the focus has been on locating critical nodes of nonlinearity within the dynamics of large-scale social systems. However, nonlinearity is not limited to any one venue. Indeed, the primary presupposition of this entire volume is that virtually all human behavior, be it aggregate or individual, is essentially nonlinear in its nature and that the linear techniques that have traditionally been used to examine such behavior are a consequence more of mathematical convenience than of theoretical necessity. This chapter takes clear aim at identifying nonlinearities inherent with individual-level behavior. However, the goal of the analysis is not to demonstrate that nonlinearities exist, but to identify the complicated structure of these nonlinearities, and to identify substantively how much we miss when we ignore them.

The Substantive Problem

The current analysis addresses two discrete literatures in political science. One body of literature is premised on the assumption that voters act rationally with regard to their candidate preferences, implying that voters are individually responsible for their own electoral behavior. Downs's 1957 work is an elegant representative example of this intellectual tradition. Yet another body of literature claims that voters are influenced in their decision-making processes by their social and political contexts. In this second case, individuals are seen as being psychologically conditioned by their environment to process information differentially in correspondence with the norms of the society within which the individuals are socially embedded. This latter case embraces a more stimulus-response view of political behavior than in the former case which identifies individual rationality as the primary motivating engine.

The theoretical situation is not, of course, one in which we are forced to choose between one of two dichotomous alternatives, rational-actor versus stimulus-response. The reality of individual behavior and attitudinal formation is most probably a blend of both worlds, and, indeed, this essay addresses the question of how these two processes may mutually interact.

With regard to rationality, it has long been suspected that individuals may not make choices objectively. Some authors suggest that part of the basis of irrational decision making can be found in Simon's early concept of "bounded rationality" (Kahneman and Tversky 1982). But the problem is difficult to sort out since one can always develop a decision rule "after the fact" that allows for virtually any decision-making outcome. Thus, the question of just how much intellectual autonomy individuals have in making their choices can be avoided if one is sufficiently clever in developing new rules that fit the observed behavior.

The rationality problem is further complicated since rational choice models are rarely isomorphically transparent with respect to the linear statistical models that are often used to support them. From a gross level of observation, it can appear that some rational-choice models may be overly complicated while their evidence-gathering statistical models are overly simplistic. That the algebraic forms of the two do not always match leaves one wondering if other processes may be occurring that are absent in the formalism which the statistical models are incapable of revealing. At some point it seems worthwhile to reduce the complexity of an individual decision-making model if one can develop a more parsimonious model that (1) embraces other aspects of social theory that may, at least partially, account for some of the observed human behavior, and (2) allows for the estimation of the total influences directly, that is, using an isomorphic algebraic form in the estimation process that structurally identifies all of the different aspects of theory.

In part due to the failure of rational-choice concepts to parsimoniously account for a great deal of human behavior that does not appear to be the product of autonomous individual action or thought, another literature—the literature on the influence of social context—has developed that fills this void. The two literatures are still quite disparate, and there is no clear view of a future common meeting ground despite some very useful attempts to bridge the gap between the literatures by identifying the differences as variations in emphasis within the broader rubric of voting models (e.g., Grofman 1987).

The contextual literature is now quite broad and still growing. While early studies were hindered by the use of cross-sectional data (or, more accurately, the unavailability of dynamically rich contextual data sets), more recent analyses have employed research designs that have both dynamic and contextual components (e.g., MacKuen and Brown 1987). Two such studies, both conducted by Huckfeldt and Sprague (1987, 1988) are of particular relevance to the analyses presented in this chapter. Both of these studies rely on survey data in which contextual information for each of the respondents is included with a rich collection of individual-level data. The data also contain repeated temporal measures across a political campaign. However, one of the primary contributions of these analyses is the authors' attempts to untangle the relative strengths of individual versus social inputs into the process of attitudinal formation. Thus, the authors attempt to determine the degree to which political attitudes and behaviors are a function of individual decision making—in a rational-choice fashion—as compared with responses to the net influence of a set of social interactions, thereby determining the limits of individual intellectual autonomy within clearly identified political and social contexts.[1]

My own analysis presented here extends the arguments made by Huckfeldt and Sprague by addressing the extent to which this rational/contextual balance is fundamentally nonlinear. Indeed, I demonstrate that the dual processes of individually rational decision making and the interactions of context can produce extremely nonlinear dynamics. Empirically, this investigation identifies a catastrophe—classically defined in the literature of nonlinear dynamics—as a potential part of these dynamics. Substantively, the model identifies situations in which individuals can experience very rapid shifts in their feelings toward a particular candidate while simultaneously experiencing relatively small changes in their feelings toward that candidate's party and the opposing candidate. Moreover, these nonlinear dynamics appear most noticeably in a theoretically anticipated political context as is explained more fully below.

The Problem of Data

The rarity of data that can be used to examine the dynamic relationships between individuals and their context is a nontrivial problem. Indeed, only a few such data sets exist at this time. Early research that examined the influence of context on individual attitudes and behaviors almost always specified the relationship in a linear fashion. Part of this was due to the fact that contextual studies have typically relied on cross-sectional data in which the prospect of finding longitudinal nonlinearity was not even possible (Miller 1956). Thus, the question of dynamics could not be addressed except in a theoretical fashion (e.g., McPhee and Smith 1962).

Some studies did use panel data that were combined with useful contextual information. Berelson, Lazarsfeld, and McPhee's 1954 study of Elmira is seminal in this regard. But such research designs are expensive to conduct, and thus are not common. They not only require contextual-level information, but also repeated measures of at least the individual-level information. Perhaps for a variety of historical reasons, the bias in the survey literature clearly is in the direction of individual-level analyses, giving evidence to the dominance of an individual-level paradigm in the thinking of many, and perhaps most, social scientists. From such a perspective, individual-level characteristics and attitudes determine other individual-level attitudes and behaviors, as if a person's mind can be understood as a self-contained unit.

However, there is a particularly rich and publicly available set of survey data that does combine individual and contextual levels of information together with repeated temporal measures of attitudinal change. This chapter's analyses are based on these data, the 1980 American National Election Study (ANES) panel study. In this study, political attitudes were elicited from the respondents in three waves (together with a November follow-up) in January, June, and September of 1980. The campaign between Jimmy Carter and Ronald Reagan thus serves as a valuable window into the mechanics of the interactions between individual- and contextual-level influences on political behavior. While these data have been studied elsewhere to identify dynamic characteristics of contextual processes (MacKuen and Brown 1987; Eulau and Rothenberg 1986), this study uses these data to investigate nonlinear aspects to these dynamics that are generally new to the literature on context.

There are three primary variables used in this study. Each are constructed from feeling thermometers that have been rescaled to range from 0 to 1. Neutral feelings are at 0.5, whereas warmer feelings are higher and cooler feelings are lower. The three variables are change in feelings for Carter, Reagan, and the Democratic party. The primary variable of interest in this study is change in feelings for President Carter during the 1980 presidential campaign between January and September.

The purpose of this analysis is not to specify fully all of the various variable inputs that may cause change in feelings for Carter. Indeed, I have presented research along these lines elsewhere (MacKuen and Brown 1987). Rather, this analysis focuses on a nonlinear specification of the interactions among a smaller subset of possible variables. Expanding the number of variables involved in these interactions simply increases the degree of nonlinear complexity, and the choice is made here to err on the side of simplicity in order to demonstrate some basic points with tractable and theoretically understandable algebraic specifications.

With regard to the other variables, it is natural for individuals to have cooler feelings toward Carter as their feelings toward Reagan improve due to the campaign coverage that the opposition candidate obtains during the year. Thus, change in feelings for Reagan is a natural variable to include in this setting.

However, it is necessary to measure another aspect of the voters' political awareness. The year 1980 was not an easy one for an incumbent president. Inflation was a bit high due to a restructuring of the oil trade (although relatively low on international standards). Moreover, the hostage crisis in Iran occupied much of the president's time and sapped much of his prestige. Thus, we want to include a more general measure of how the voters evaluated the Democratic approach to government than their feelings for the president. Indeed, we want to know how their changes in feelings for the Democratic party in general changed in correspondence with changes in their feelings for Carter and Reagan.

Briefly, 1980 was a year in which one might expect an overall decline in voters' feelings for the Democratic party. At some point in time during this gradual decline, I hypothesize here, some voters will "throw in the towel," so to speak, and more rapidly abandon their support for Carter. At this point (call it a *tipping point*) one would expect a rapid drop in voter feelings for Carter combined with modest changes in the feelings for Reagan and feelings for the Democratic party. This expectation is a nonlinear relationship between change in feelings for Carter and change in feelings for Reagan and the Democratic party and is based on theoretical motivations underlying the construction of the model investigated here.

Contextual variables are also introduced, later in this analysis. These variables are county-level proportions of support for the Democratic and

Republican presidential candidates in November of 1980. These variables are used to test a specific theory relating to the dependence of the mutual dynamics between the three attitudinal variables mentioned above with respect to social context. This theory addresses the idea that tipping points arising from nonlinear models of human behavior may be contextually dependent. The theoretical connection between such nonlinear dynamics and social context is not new (e.g., Huckfeldt 1980; Schelling 1978). However, the empirical finding here that extreme forms of nonlinearity may be contextually dependent in a theoretically anticipated fashion significantly adds to our understanding of the role of context in shaping individual attitudes.

Preliminaries: A Linear Regression Model

The mechanics of nonlinear modeling are not as straightforward as those of linear modeling. In general, estimates for a linear regression model can be obtained in a matter of seconds, even in situations with large samples. Nonlinear models require more work, both in terms of human and computer effort, and typically it is difficult to obtain reliable estimates for a nonlinear model using nonlinear least squares in a period of less than a few months at least. Thus, one needs to be persuaded that what one receives from all of this effort is worth the total cost, human and otherwise.

To demonstrate how valuable the effort can be, a useful heuristic strategy is to examine the results of a linear regression model that uses the exact same variables as those that are used in the nonlinear model. Improvements in fit as well as descriptive richness of the nonlinear model can then be evaluated from this linear baseline. This is the strategy pursued here. Initially, only three variables are used in the comparison: changes in feelings for Carter, Reagan, and the Democratic party. In this way, improvement in the models can be isolated as we shift from a traditional linear approach to the more complex nonlinear specification.

Table 3.1 presents the results of a linear regression in which change in feelings for Carter between January and September of 1980 is a function of change in feelings for Reagan and change in feelings for the Democratic party. First, note that the fit of the model is not very high. Note also that the parameter estimate for change in feelings for Reagan is positive, encouraging the initial interpretation that voters who tended to feel more warmly for Reagan rated Carter more highly. Under such a scenario, change—positive or negative—influences the evaluations for each candidate in unison. However, note that the estimate for change in feelings for the Democratic party is three times higher than that for Reagan. This is as would be expected given the closer substantive correspondence between the incumbent president and his party. However, the positive estimate for change in feelings for Reagan,

TABLE 3.1. Change in Feelings for Carter without Partisan Identification, January–September 1980

R^2	0.0842
Adjusted R^2	0.0799

	Parameter Estimate	Standard Error	Prob > |T|
Intercept	-0.081	0.011	0.0001
Reagan	0.097	0.042	0.0211
Democratic party	0.324	0.058	0.0001

$N = 427$

Note: All variables are feeling thermometers with a range of 0 to 1.

while interpretable, still remains a bit of a puzzle. We will return to this puzzle later.

Table 3.2 is similar to table 3.1 with the exception that the voters' partisan identification is added to the specification. This is done simply to test the stability of the estimates presented in table 3.1 that employ the modestly more meager specification. With partisan identification, the positive estimate for change in feelings for Reagan re-appears with approximately the same

TABLE 3.2. Change in Feelings for Carter with Partisan Identification, January–September 1980

R^2	0.1218
Adjusted R^2	0.1156

	Parameter Estimate	Standard Error	Prob > |T|
Intercept	-0.003	0.021	0.9031
Reagan	0.116	0.041	0.0054
Democratic Party	0.349	0.057	0.0001
Partisan ID	-0.021	0.005	0.0001

$N = 427$

Note: All variables except partisan identification are feeling thermometers with a range of 0 to 1.

magnitude. The estimate for change in feelings for the Democratic party also remains stable, and the overall fit of the model improves somewhat. Thus, our baseline model for this analysis is that presented in table 3.1, a spartan but generally stable set of linear estimates relating change in feelings for Reagan and the Democratic party to change in feelings toward Carter.

The Nonlinear Model

In its simplest form, the fundamental proposition in this analysis is that gradual change in feelings for Reagan and the Democratic party will, at some point, result in a larger—and somewhat sudden—change in feelings for Carter. This will occur as voters reach some threshold level of support that can be identified among the three variables. The problem is to identify a specification of this process using individual-level information that can be found in the available survey data. Contextual information plays a crucial role in the test of this proposition, and its use is explained more thoroughly below.

I begin the specification of the model by describing change in feelings for Reagan and the Democratic party. For economy (an important consideration given the demand on computational resources for highly nonlinear models), I describe change in these two variables as simply as possible, reserving for later a more challenging nonlinear specification for change in feelings for Carter. Here, change in feelings for Reagan and the Democratic party is written as a linear function of previous values of these variables. Since we are dealing with continuous time between two time points, note that the functional linearity of this part of the specification still allows for a high degree of longitudinal nonlinearity along the individual variable trajectories. Thus, we have

$$dR/dt = p_1 + (p_2R) \tag{3.1}$$

$$dL/dt = p_3 + (p_4L) \tag{3.2}$$

In equations 3.1 and 3.2, p_i are parameters that control the rate of growth of the two variables. R is the variable describing feelings for Reagan, whereas L represents feelings for the Democratic party.

To describe change in feelings for Carter, I develop a theory that is based on a variety of interacting components. To begin, in a rational world one would expect feelings for Carter to be highest when feelings for the Democratic party are also highest and feelings for Reagan are the lowest. Moreover, warmth in feelings for Reagan should decrease the strength of the warm feelings for Carter, but more so if feelings for the Democratic party are also lower. Crucially, feelings for a candidate are much more target-specific than feelings for a party, and, in balance, feelings for candidates should dominate over feelings for a party. Thus, warm feelings for the Democratic party should help support warmer feelings for Carter, but the relationship should decay in the presence of very warm feelings for Reagan. For example, feelings for Carter should be less than the maximum when a voter feels very warm for Reagan, despite potentially very warm feelings for the Democratic party. In this example, warm feelings for the Democratic party can only partially buffer losses in warmth for Carter in the presence of very warm feelings for Reagan. On the other hand, the office of the president does have some advantages for the incumbent. For example, the president has continual access to the media. The office itself also can act as a protective shield against extremely negative feelings—say, due to a sense of obligatory patriotism toward the office and whoever holds that office, especially in the absence of an attractive alternative candidate. Thus, voters that feel very negative toward Reagan and the Democratic party may very likely feel somewhat warmer toward Carter.

This balancing act between the feelings for the two candidates and the Democratic party can be represented algebraically as follows:

$$W_1 = (1 + L - R)/2 \tag{3.3}$$

Here, W_1 represents the attitudinal support for Carter in a rational world in which voters allocate their support based on their feelings toward the incumbent's party and the opposition candidate. Note that this is a static quantity, and thus it remains a fixed feature in a specification of change in feelings for Carter.

The relationship identified in equation 3.3 and described verbally above is graphically described in figure 3.1. In this figure, feelings for Carter are represented on the vertical axis whereas feelings for Reagan and the Democratic party are represented on the floor axes. From this figure, note that feelings for Carter are highest when feelings for the Democratic party are very warm and feelings for Reagan are low. Increases in warmth for Reagan decrease support for Carter, but warm feelings for the Democratic party buffer this loss.

Initially, change in feelings for Carter should be described in terms of distance from the plane in figure 3.1. Thus, the plane should represent an attractor based on rational expectations. (Further on in this analysis the plane is parameterized to allow the estimation procedure to determine its exact angle and placement.) If someone's feelings for Carter are warmer than the planar attractor would suggest is appropriate given the person's feelings for Reagan and the Democratic party, then the expectation will be that the person's feelings for Carter will begin to decay, and thus approach a point on the attractor. Similarly, if someone's feelings for Carter were cooler than the planar attrac-

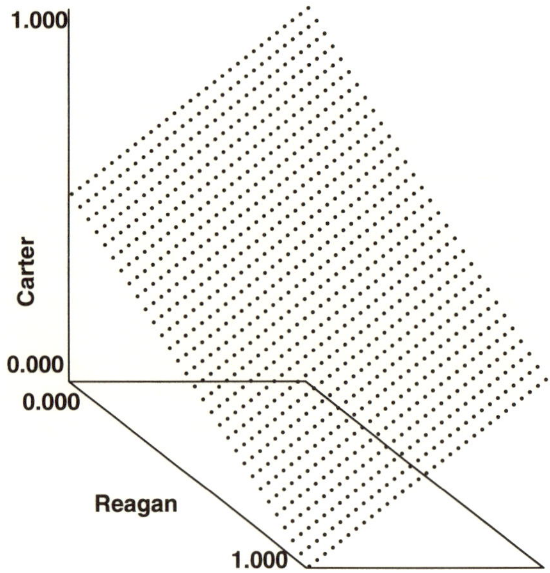

Democratic Party range = [0.000, 1.000]

Fig. 3.1. Feelings for Carter based on static rational expectations

tor would indicate under those conditions, then their feelings for Carter would be expected to improve as the campaign year progresses. These expected movements in the trajectory of feelings for Carter can be captured algebraically as

$$dC/dt = W_1 - C. \tag{3.4}$$

Here, C represents feelings for Carter. When C is below W_1, change in C is in a positive direction (i.e., increased warmth for Carter). When C is above W_1, change is toward cooler feelings.

However, a person's ideal attitudes (i.e., as represented on the system's attractor) are not likely to be independent of their own feelings for Carter. Thus, we wish to express the attractor in terms of C as well as W_1. This helps express the concept of endogenous feedback in the attractor and abandons the notion that the attractor needs to be exogenously determined. Thus, these ideal points will likely depend on the current state of voters' feelings for

Carter. Moreover, there is likely to be a limit to the level of change that most people will experience (i.e., bounds on the attractor that are based on feelings for Carter).

One way to describe such characteristics algebraically is to say that the attractor will include a logistic expression based on feelings for Carter. Thus, the attractor's value for any given combination of feelings for Reagan and the Democratic party will be a function of the current value of C as well as the negative of the value of C raised to some power. Raising C to some power gives the model the polynomial flexibility to capture the idea of a voter abandoning support for Carter quickly after reaching some threshold in terms of feelings for Reagan and the Democratic party. However, this cannot be accomplished with a squared value of C. The typical practice of squaring C when forming a logistic structure is somewhat arbitrary, and there is no reason not to use higher powers should there be a theoretical expectation that the approach to the limit needs to be given greater definition.

The theoretically anticipated dynamics presented here (i.e., rapid change in one variable in correspondence with gradual change in others) requires the disappearance of an attractor for a given trajectory and the sudden appearance of an alternative attractor (other than infinity). From a substantive perspective, the tipping point concept mentioned previously is crucial. At a point in time during the campaign—and based on contextual and historical voter contexts—some voters will reach a threshold at which they will rapidly abandon their support for Carter (e.g., "throw in the towel") and psychologically rush to support Reagan. I develop this idea more thoroughly below in the discussion of the influence of social contexts, but the basic concept is that voters will be held to an allegiance to Carter due to pressures emanating from their surrounding milieu. Initially, their feelings for Reagan can warm, but their feelings for Carter will not correspondingly cool. There will be a delay caused by an individual's adherence to community and group norms. This delay may not last forever, however. Eventually an individual who continues to warm to the Reagan appeal will break away from the pressures of friends, family, neighbors, and co-workers. When this breakage occurs, the system will be dominated by a different attractor, and the influence of the first attractor (allegiance to Carter) will quickly wane.

From a modeling perspective in the current setting, this requires a power of three or greater. Squaring C will not allow for an alternative non-infinity attractor to exist in the model's phase space.[2] A cubic is the minimum degree polynomial structure that can produce a bimodal attractor structure in the given situation. Moreover, while higher degree polynomial structures can produce more complex attractor topologies, I have not yet developed theoretical justifications for the use of such forms, and the choice is made here to follow a more minimalist specification that can be fully understood and justified using substantive arguments. Thus, in this model, C is raised to the power three.[3] All of this can be expressed using the phrasing

$$dC/dt = (C - C^3 + W_1) - C. \tag{3.5}$$

At this point the model needs to be parameterized in order to allow for empirical flexibility. This parameterization also eliminates the apparent algebraic canceling of the two uncubed appearances of the variable C in equation 3.5. The parameterized model can be expressed as

$$dC/dt = p_8[(p_5(C - p_6) - (C - p_6)^3 + p_7W_1) - C]. \tag{3.6}$$

In this expression, parameters p_5 and p_7 act to scale the values of the state variables, thereby controlling the shape of the attractor. However, the parameter p_6 plays a different role. There is no reason to suspect that the nonlinearity of the influence of C in the structure of the attracting surface will be symmetrically bounded by its own range between zero and one. From a numerical point of view, some of the more extreme nonlinearities in the attracting surface may occur outside of the range of C, leaving only residual nonlinear traces within the realistic boundaries of this variable. Where the nonlinearities occur in the hypersurface is an empirical question that must be resolved with the estimation of the parameters. This latter parameter allows the estimation program to determine the exact vertical placement of the polynomial nonlinearities in the attracting surface. (Indeed, the empirical results presented below suggest that the placement of these nonlinearities is highly dependent on a voter's political context.) The parameter p_8 controls the speed with which voter trajectories for C approach the attracting surface.

Nonlinearities in the attracting equilibrium surface (determined by setting $dC/dt = 0$ and solving for the equilibrium values of C), are now a result of the interplay between C and C^3. Functionally, the interplay is uniformly the same across the entire range of R and L. However, it is likely that the interplay will vary depending on the values of these variables. Within some range of the variables, the interplay may result in strong nonlinearities in the equilibrium surface. However, outside of that range, the nonlinearities may be more moderate. It is not possible to determine in advance of the estimation of the parameters where the nonlinear range is located. But the model must be sufficiently flexible to allow the data to determine the lateral placement of the nonlinearities. Moreover, it is necessary to establish expectations with regard to the key inputs that lead to their appearance.

To do this, an additional measure is needed. The measure will be used to "turn off" the interplay between C and C^3 (and thus the severity of the nonlinearities in the equilibrium surface) above and below a certain range of values for R and L. This measure needs to collapse the joint movements between these two latter variables into one dimension. The new measure needs to be largest when both R and L are large, and smallest when both of these variables are small. This new measure is

$$W_2 = (R + L)/2. \tag{3.7}$$

This new measure is included in the model for change in feelings for Carter as in equation 3.8.

$$\begin{aligned} dC/dt = p_8[&p_5(p_9 + W_2)(1 - p_{10}W_2)(C - p_6) \\ &- (C - p_6)^3 + (p_7W_1) - C] \end{aligned} \tag{3.8}$$

In equation 3.8, the term $p_5(p_9 + W_2)(1 - p_{10}W_2)(C - p_6)$ tends toward zero when W_2 is near some lower limit (when $p_9 + W_2 = 0$), or when W_2 approaches an upper limit (where $1 - p_{10}W_2 = 0$). The parameters p_9 and p_{10} allow these limits to be determined empirically. In either of these situations, the interplay between C and C^3 decreases and the severity of polynomial nonlinearity on the equilibrium surface is lessened.

One final change in the model is needed. The nonlinearities in the equilibrium surface should reflect the influence of context. While the system as currently specified allows for a great deal of flexibility in determining where the nonlinearities appear, it does not yet allow for the model to reveal these aspects differentially across varying political contexts. To do this, we need to condition the system's parameters with respect to a measure of context. In this analysis, two such measures are used. They are the proportions of the eligible population in each respondent's county of residence that voted for the Democratic presidential candidate, and similarly the Republican candidate, in 1980. These proportions reflect the partisan flavor of each respondent's surrounding environment.

To condition the system for context, each parameter is written as a linear function of a contextual variable. Following the established practice used elsewhere in this volume (see also Brown 1991), and using one parameter as an example here, the parameters are now estimated as $p_1 = p_{1A} + p_{1B}CONTEXT$, where $CONTEXT$ is, say, the proportion of the eligible population that votes Democratic.

Conditioning the parameters is crucial in the current setting for substantive reasons. Contextual influences should affect the movement of the trajectories in the model. This influence should be apparent in the manifestation of nonlinearities in the equilibrium surface. Briefly, the equilibrium surface should "pull" toward the pro-Carter end of C when the local political context is Democratic. That is, the norms of the local environment should cause resistance to a decay in Carter feelings when an individual's other feelings for Reagan get warmer and their feelings for the Democratic party get cooler.

Recall that the expectations of a rational actor would be to modify his or her feelings in the direction of the attracting surface shown in figure 3.1. However, the contextual literature has reported repeatedly that an individual's surrounding social environment acts to maintain community norms by directly and indirectly pressuring deviants to conform to the more widely held views. Thus, if a respondent's normal state based on the relative balance of feelings for Carter, Reagan, and the Republican party would be to feel cooler toward Carter, a Democratic environment would act to delay that change until the psychological pressure within the individual (due to further changes in his or her evaluations of Reagan and the Democratic party) passed some critical threshold. This would result in a rapid psychological departure from the neighborhood norms and a return to the more individualized balance between the three feelings (i.e., the reemergence of the rational-actor model).

The Results

The entire system of interdependent equations can now be represented as equations 3.1, 3.2, and 3.8. Essentially due to equation 3.8, the system is highly nonlinear in both parameters and states. As with many of the other models presented in this book, estimating this system is a nontrivial problem requiring a nonlinear least squares approach. The parameter estimates for the entire system, including the conditioned estimates, are presented in table 3.3.

The primary interpretive lessons to be drawn from the estimated system are best obtained from the graphical analysis that follows. However, some important points are observed directly from table 3.3. First, note that the fit of the model for change in feelings for Carter is much higher (approximately four times higher) than that obtained using a linear regression model with the same variables as presented in table 3.1. Also, note that all of the parameters have acceptable signs. The only estimate with a negative sign for change in Carter feelings (equation 3.8) is p_9, which is used to help define the region of more severe nonlinearity in the predictive hypersurface for the model. The negative value simply indicates that this region is slightly skewed across the range of values of W_2. However, the degree of asymmetry is slight since the magnitude of this parameter is small, and the estimate does not test significantly different from zero. The Simon F statistic indicates the change in the fit for the overall system when the given unconditioned parameter estimate is fixed at zero as compared with its estimated optimal value, thereby giving an indication of the relative importance of the parameter to the model.

TABLE 3.3 Parameter Estimates

Parameters	National Estimates	Chi-Square ($df = 1$)	Simon *F*	Democratic Context	Republican Context
p_1	0.26724	177.668	0.26446	-0.014208	0.03741
p_2	-0.38401	84.316	0.30368	-0.030225	-0.00415
p_3	0.19066	2.496	0.21388	-0.019417	0.02407
p_4	-0.38721	3.519	0.62810	0.029381	-0.02916
p_5	5.98186	9.983	0.01546	0.001897	-0.00334
p_6	0.83441	24.372	0.05422	0.008622	-0.04221
p_7	0.88157	95.546	0.17592	0.052890	-0.09321
p_8	0.41774	59.310	0.10373	0.066099	-0.06038
p_9	-0.13497	0.351	0.00916	0.078435	-0.11694
p_{10}	1.24597	23.406	0.63444	0.043908	-0.01458

Equation	National (unconditioned) Fits	Democratic Fits	Republican Fits
Reagan	0.22963	0.24074	0.24292
Democratic party	0.28320	0.28387	0.28408
Carter	0.31120	0.31979	0.32424
System average	0.27468	0.28147	0.28375

Graphical techniques help simplify the interpretation of models with a high degree of functional nonlinearity such as the one under investigation here. The most basic approach is to determine the extent to which the nonlinear specification alters the rational-actor equilibrium surface presented in figure 3.1. Figure 3.2 displays such a surface using unconditioned (i.e., national-level) parameter estimates.[4]

In figure 3.2, note that the equilibrium surface, while nonlinear, does not depart strongly from the planar surface of figure 3.1. Interpretively, this suggests, in the absence of information about an individual's context, that voters appear to be following a rational-actor's approach to candidate evaluation. The conditionality of this last statement is critically important. Most political surveys do not contain information about context. Thus, *in the absence of this information*, social scientists can find evidence that confirms an individual-level approach to the study of politics using more traditional linear models, since that is all that would be needed to identify the relationship presented in figure 3.1. The falseness of this view becomes apparent when the model is conditioned with respect to a voter's political environment, as is

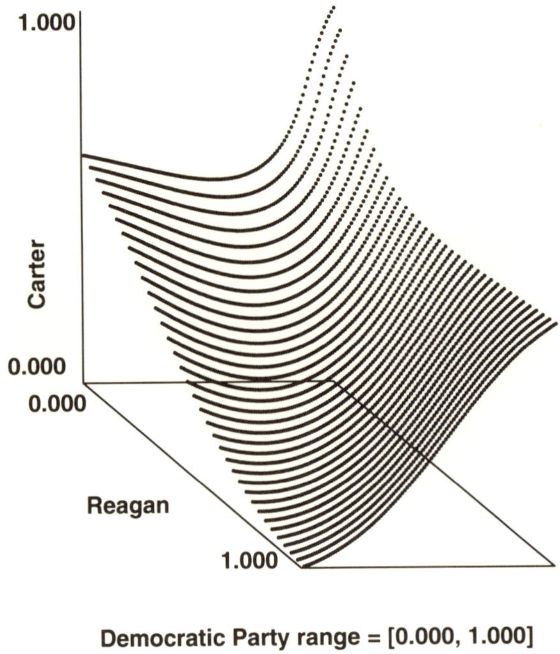

Fig. 3.2. Feeling thermometer equilibrium surface

demonstrated below. First, however, it is useful to show how individual-level attitudinal change corresponds with the equilibrium surface shown in figure 3.2.

Figure 3.3 presents a phase diagram that is a cross section of the equilibrium surface in figure 3.2. To construct this cross section, feelings for Reagan were fixed at a low level (0.1). The horizontal axis represents feelings for the Democratic party. The equilibrium surface is now represented as the nearly horizontal line (curving up slightly, moving from left to right) that cuts across the figure. The other dotted curves represent sample trajectories for individual voters. In this figure, the sample trajectories are extended to allow the reader to observe clearly the directional movement of the trajectories toward their point of attraction. Note that there is a global attractor on the equilibrium surface in this figure. Moreover, note that movements toward this attractor generally have only moderate amounts of vertical movement, suggesting the absence of sudden and large-scale change in feelings for Carter in the fashion of a catastrophe, classically defined.

The simplicity of the equilibrium surface of figure 3.2 changes dramat-

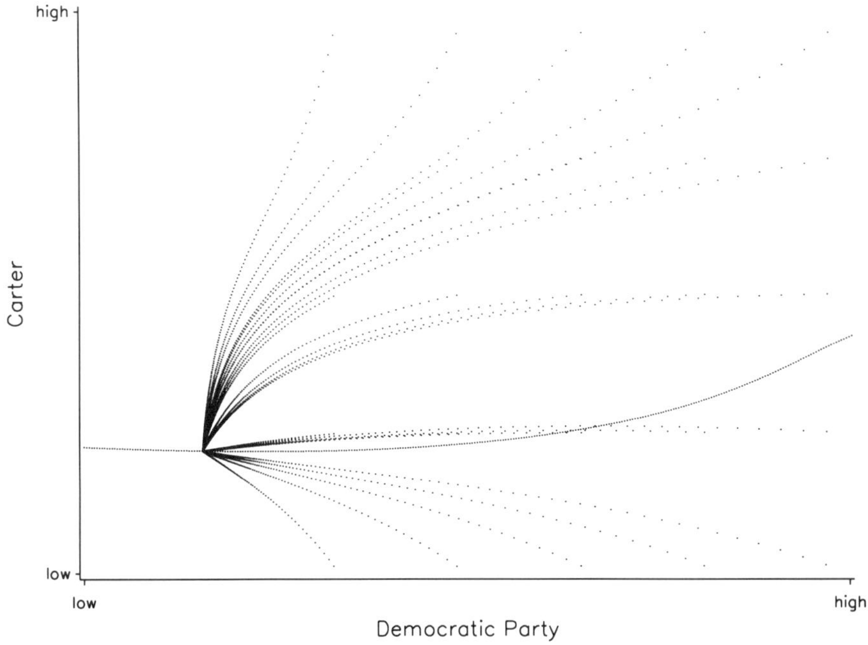

Fig. 3.3. Carter and Democratic party feelings, National condition

ically when we condition for an individual's political context. Figure 3.4 presents the equilibrium surface for the condition in which a voter's general environment is strongly democratic. In this figure, note that the equilibrium surface has a large bulge in areas in which feelings for the Democratic party are warm and feelings for Reagan are cool or moderate. This is precisely what is expected with regard to the influence of context on the formation of attitudes among individuals. When voters live in strongly Democratic areas, these areas influence voters to depart from that which would otherwise be their own individualistic evaluation of the candidates. This departure is seen in terms of the equilibrium surface "pulling" toward warmer feelings for Carter than would occur otherwise given the rational-actor model.

However, figure 3.4 is complex. The pull toward the norms of the Democratic environment does not happen across the entire figure. Indeed, the pull is basically isolated in situations in which feelings for the Democratic party are warm and feelings for Reagan are cool or moderate. This suggests that the pull of the political environment is greatest when feelings toward a candidate that opposes the norms of that environment are not particularly

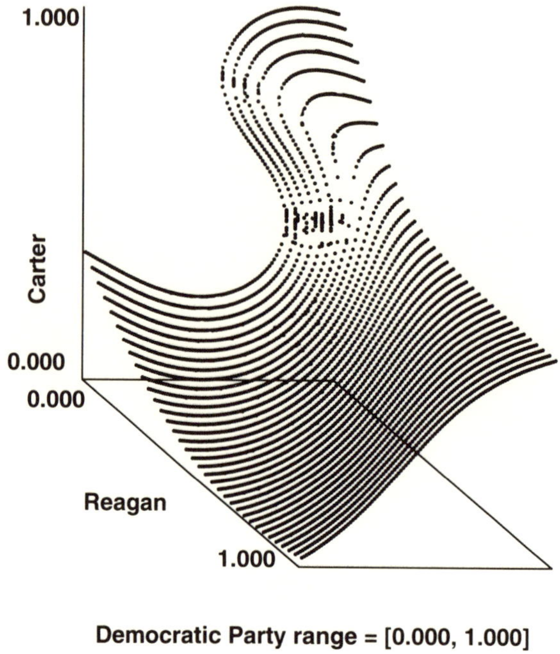

Fig. 3.4. Equilibrium surface, Democratic context

strong. Put differently, and as measured in the current setting, intensely felt attitudes can, at least partially, immunize the voters from the influence of context.

When comparing figures 3.2 and 3.4, note that the lower part of the surface in figure 3.4 (the area of cooler feelings toward the Democratic party) is much lower than the comparable area in figure 3.2. Thus, the norms of the Democratic party pull feelings for Carter upward. But once a voter makes a psychological departure from these societal Democratic norms, the hostility toward Carter is greater than that which would otherwise be expected. This result finds correspondence with research by Huckfeldt and Sprague (1988, 1987) with regard to the behavior of individuals who hold minority views relative to their local context. The comparison of figures 3.2 and 3.4 also demonstrates both the nonlinearity and the strength of this polarizing contextual influence.

The bulge in figure 3.4 represents a multidimensional catastrophe. Figure 3.5 shows one dimension of this clearly by presenting a cross section of the equilibrium surface of figure 3.4 using a low value of support for Reagan. The S-shaped curve is the equilibrium surface, whereas the other dotted

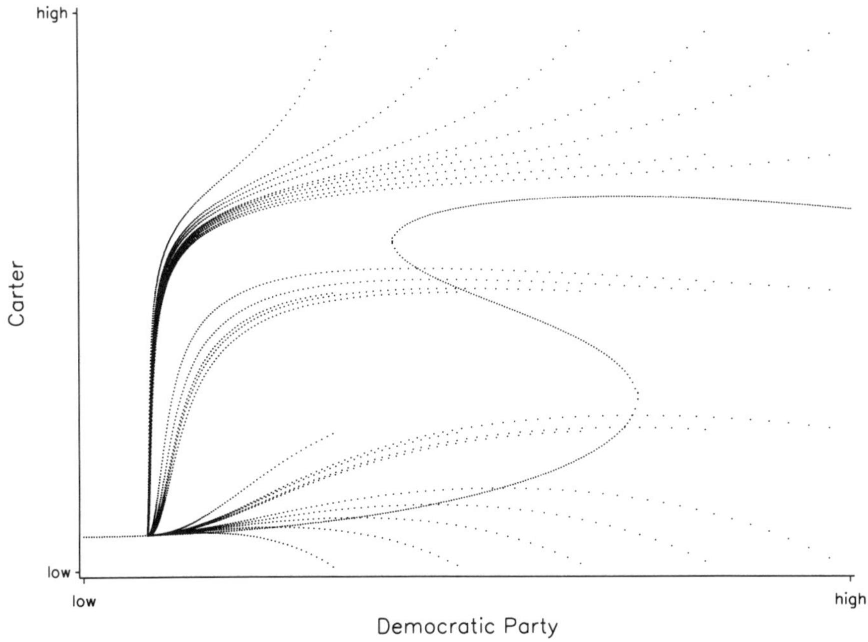

Fig. 3.5. Carter and Democratic party feelings, Democratic environment

curves represent sample trajectories for voters. In this figure, note that there is a great deal of vertical movement in many of the trajectories, especially those with starting conditions having high support for Carter. Thus, in certain areas along the horizontal axis, small change in feelings for the Democratic party correspond with large and rapid changes in feelings for Carter.

The catastrophe potential of this model is not limited to change in feelings for the Democratic party. The multidimensional aspect of the catastrophe is also evident with regard to change in feelings for Reagan. Figure 3.6 presents an alternative cross-sectional view of the equilibrium surface in figure 3.4. In figure 3.6, feelings for the Democratic party are held at a relatively high level (0.8). The reverse S-shape in this figure is the equilibrium surface. The sample trajectories move toward a global attractor on this surface. As with figure 3.5, in some areas on the horizontal axis of figure 3.6, small change in feelings for Reagan correspond with large decays in feelings for Carter. Moreover, many trajectories are drawn toward this area of catastrophe potential. The upper end of the equilibrium surface again delays the downward movement of the trajectories until sufficient support for Reagan builds. Again, this happens only in situations in which voters live within a

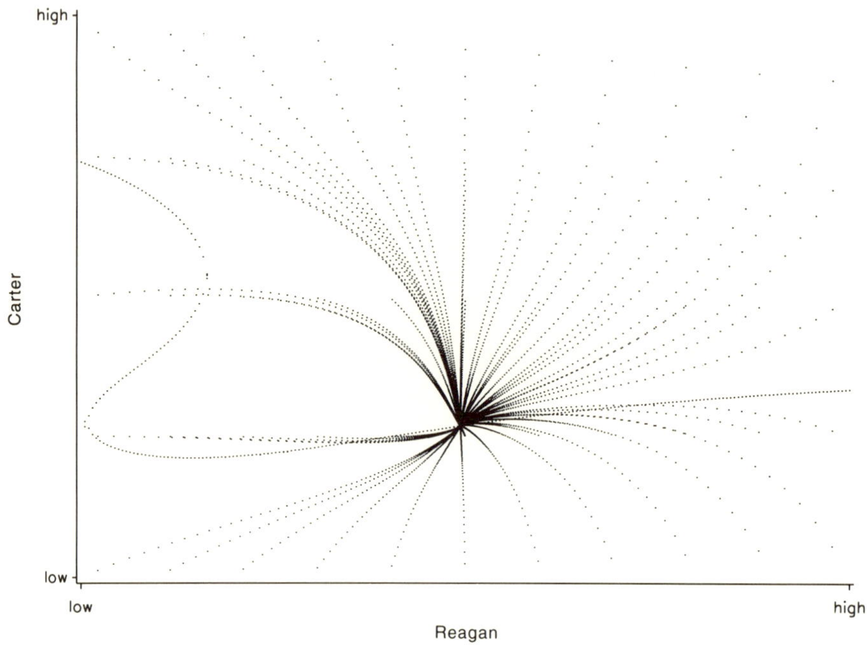

Fig. 3.6. Carter and Reagan feelings, Democratic environment

predominantly Democratic milieu, and in which the voters' feelings for Reagan are low to moderate and their feelings for the Democratic party are high.

Switching to a Republican political environment completely eliminates the bulge in figure 3.4, as would be expected according to theory since change in feelings for the Democratic candidate should be most affected in a positive sense by a Democratic political environment. Figure 3.7 displays the equilibrium surface for the model within a Republican environment. The bulge in the surface is absent because the environment no longer delays the decay of feelings for Carter by influencing the voter to conform to community norms. However, we do not have a simple return of the rational-actor model. It is with this figure that the multifaceted influence of context becomes apparent. While there is no contextually determined severe nonlinearity in the equilibrium surface, there is nonetheless a dramatic change in the surface as compared with that of figure 3.4. In a Republican environment, the equilibrium surface is dramatically cooler with regard to feelings for Carter. This is so across the entire range of feelings for the Democratic party, but it is particularly apparent in situations in which feelings for the Democratic party are warm. In figure 3.7, feelings for Carter are relatively uniform across the range

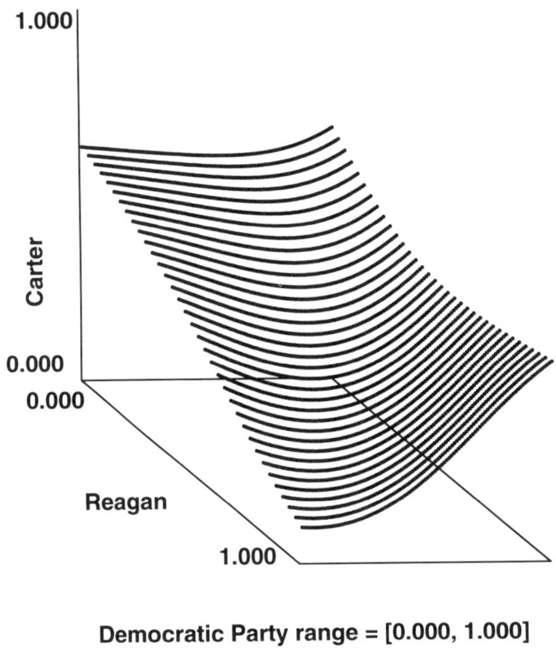

Fig. 3.7. Equilibrium surface, Republican context

of feelings for the Democratic party. Even when feelings for Reagan are very low and feelings for the Democratic party are high, there is no large increase in feelings for Carter as is apparent in figure 3.4 for the nation more generally. Thus, political context not only influences nonlinear characteristics of the attracting surface, but also the vertical placement of the surface as well.

Substantively, this implies that political context influences voters' attitudes in a highly complex fashion. The influence is always to pull the voters' attitudes in the direction of the community norms. However, sometimes the influence cuts across the entire range of the variables, thereby generalizing the effect on the perception of a candidate. But in other situations the influence of context is greatly complicated. Severe forms of nonlinearity can appear in the attracting equilibrium surfaces while the overall perceptions can be simultaneously affected more generally across the range of the variables as well.

Discussion

The primary results of this analysis are both substantive and theoretical. But since the substance and the theory are directly associated with the nonlinear algebra investigated here, it is useful to begin the more general discussion with some comments regarding the tradition of using linear models in most social scientific analysis. Throughout these comments it is useful to remind oneself that I am not making a negative argument criticizing other people's use of linear models, since I too use linear approaches in appropriate settings. Rather, I am making a positive argument for a more sympathetic consideration of nonlinear algebraic forms, especially with regard to modeling politics and social structure over time.

Remember that linear regression is a model, not a method. The model is an artifact of relatively simple optimization calculations in which the functional form of an estimator for a slope can be readily derived and evaluated from the linear algebra. The general linear model can be very useful in identifying correlational evidence with respect to changes among a set of variables. The exact same model can be used in a wide variety of situations since there is no implicit nor explicit social theory underlying the model. The association with social theory is in the choice of independent variables. Thus, in a very real sense, the linear model is a model of convenience, not of theory. While this, by itself, is not necessarily a bad thing, an unfortunate consequence of its widespread use is that we too often butcher our theories to accommodate linearized computational requirements. Rather than use the general linear model as a starting point for our nonlinear investigations, we too often stop our research just at the point at which further efforts could yield great rewards.

But understand this: I am not concerned with how other people conduct their empirical analyses. My concern is the distortion that I see in the way many of us view our world and the evolution of human culture. There is nothing inherently "bad" about the general linear model. It is just that its overuse becomes a testament to our loss of understanding of the complexities of human societies, not a panaceic endorsement of the usefulness of the linear form.

The results found in table 3.1 offered both poor performance with regard to the linear model's fit and a small puzzle, for the latter of which I offered a strained interpretation. The puzzle, a positive slope for change in feelings for Carter given change in feelings for Reagan, basically vanishes in the analysis of the nonlinear model. Indeed, given the high degree of nonlinearity in the equilibrium surface for particular environments (e.g., fig. 3.4), it is easy to see how a linear model could run into trouble in finding appropriate directions in which to aim its slopes. More complex linear models (e.g., models with more independent variables) complicate the picture since the models have more flexibility among the deterministic and stochastic characteristics of the various variables to bury the underlying nonlinearity that is authentic to the data. Thus, a complex linear model can offer both an incorrect functional form and a mechanism by which evidence of this misspecification can be overlooked.

This analysis conducts a heuristic demonstration. A linear model is directly compared with a nonlinear model using the same minimal set of variables. Since the algebraic structure of the linear model is not directly tied to any particular social theory, the use of the linear model here is in the spirit of searching for correlational associations among the variables, not an atypical setting for the general use of such models in the social sciences. The nonlinear model, filled with theoretical expectations that are tied to the algebraic intricacies of the specification, both fits the data more closely and offers a far richer interpretive setting with regard to our understanding of politics.

The cost of using such a nonlinear model is not insignificant. The nonlinear component of this research took more than one year to complete. The estimation process alone (once the model specification was set) took nearly six months, running almost continuously on a supercomputer (an IBM 3090). The linear component of this research took approximately one-half hour, including the time required to set up the necessary program. Given these costs in time and effort—both human and computational, it is no wonder that the use of linear models still dominates social scientific research.

The nonlinearity of the model investigated here allows for a significant harvest of substantive insight into the influence of social context on the formation of individual attitudes. Two substantive theories addressing the motivational structure of change in attitudes are evaluated. First, the idea is examined that voters evaluate candidates from an individualistic perspective, placing their feelings for a candidate in correspondence with their feelings for the opposing candidate and the party of the incumbent. The equilibrium surface of the model using national (unconditioned) parameter estimates suggests that, on average, this model appears to correspond fairly closely with the national data, *as long as information relating to an individual's social context is omitted from the analyses*. However, when an individual's political environment is also considered, the model reveals highly nonlinear structural characteristics to the longitudinal dynamics of attitudinal change. Briefly, an individual's social environment acts to distort the individualistic evaluative processes involved in perceiving political candidates. The distortion is in the direction of the norms of the general community within which the voters live. This substantive result finds strong correspondence with the large literature regarding the influence of context on individual behavior and attitudinal formation.

The finding that nonlinearity in social dynamics is contextually dependent is new to our understanding of human society. It suggests that it is possible to conduct an analysis of a nonlinear model with respect to a body of data and not find significant levels of nonlinearity within the embedded dynamics if contextual questions are not simultaneously addressed. This emphasizes the need to examine questions of social context as an inherent component of nonlinear social change. It also adds to our improved understanding of the complexity of human societies. Nonlinear complexity must include the potential for interaction between individuals and their surrounding milieu. The current analysis only scratches the surface of the potential degree of nonlinearity that resides within the complex set of interactions that fill our daily lives. But that this nonlinearity is an essential aspect of the fabric of our existence is certain. That its structural complexity is still hardly known to us is probably equally true.

CHAPTER 4

The Anatomy of a Landslide: Johnson and Goldwater in 1964

Landslide electoral victories, for a variety of reasons, can be extraordinary political events, both in terms of their historical importance as well as their internal nonlinear complexity. Interestingly, there have been only a few scholarly attempts to isolate landslides as examples of large-scale electoral change. Notably, Kelley (1983) has made an important analysis of a number of landslides from the level of the individual voter. But there is little known of the aggregate electoral structure of such events. Typically, the voting literature notes that "so-and-so" won with a large margin, without reference to the existence or absence of highly patterned and nonlinear processes of complex social change. But such elections are not everyday events, and their mass structure deserves close examination. Indeed, this analysis demonstrates that landslides can be much more than uniform voter swings. Rather, they should be seen as examples of large-magnitude, rapid electoral change that are best characterized by potentially complex, nonlinear, and contextually conditioned dynamic social processes.[1]

This chapter's analysis examines the structure of the aggregate mass electorate during the 1964 presidential contest between Lyndon Johnson and Barry Goldwater from such a contextual point of view. The 1964 election was the largest presidential landslide election in the United States in this century (explained more fully below). A model is presented that allows for a comparative examination of a variety of contextually defined nonlinear components to the large-scale voter movements. More specifically, the model draws attention to contextual theories of aggregate partisan change that have been proposed by Huckfeldt (1983), Przeworski and Soares (1971), and Huckfeldt and Sprague (1987, 1988), as well as the influence of local partisan context on party behavior as reported by Beck (1974).

The mid-1960s were tumultuous years in the United States. The Civil Rights movement was in full bloom at that time. Debates about governmental

---

Part of this chapter is reprinted, with permission, from the article titled, "Nonlinear Transformation in a Landslide: Johnson and Goldwater in 1964," *American Journal of Political Science*, 37:582–609, No. 2, May 1993.

social programs filled the legislative halls. The cold war was hot. Moreover, the conservative and liberal movements were both organizing their electoral forces. The 1964 election was the first large-scale battle between these two forces. Indeed, the ideological struggle that emerged so forcefully in the 1964 election continues with significant intensity today. Thus, an enhanced understanding of that crucial election may help us understand some aspects of more recent electoral politics. But, more importantly, a greater knowledge of the 1964 election will help us begin to develop a better understanding of electoral landslides per se, including the future potential of rapid large-scale electoral change during times of social turmoil.[2]

In characterizing the dynamic processes of aggregate voter movements that might occur during a landslide, this analysis identifies a number of distinct structural mechanisms of change. For example, one mechanism portrays a social process in which voters from one party interact with voters from the other party, explicitly addressing theoretical concepts of contagion and diffusion. Among the other processes examined is the influence of the current dominance of one party's local support relative to that of the other political party. Additional mechanisms of mass partisan change are included in the complete specification developed below. Crucially, all such mechanisms are evaluated while controlling for the momentum of political change. This allows for an identification of accelerating and decelerating influences in the dynamic processes. For example, it may be that Democrats interacting with Republicans cause more Democrat-to-Republican conversions when there are proportionately more Democrats locally. Alternatively, the accelerated conversion process may require the "bunching" of Democrats, such that Democrats would be interacting with other Democrats, thereby gaining mutual reinforcement and consequently yielding a perceptible shift in the norms of the political environment. This would put added contextual pressure on the conversion process. The relative impacts of all of the specified mechanisms of electoral change are evaluated below.

In addition to identifying structural components of the aggregate voter movements during a landslide election as discussed above, this analysis seeks to answer a number of fundamental questions regarding the overall impact of a landslide on groups of voters. These questions can be introduced by way of an analogy involving a river. When a river flows past straight banks, all is calm. There are no whirlpools, no rapids. The movement of the water is relatively smooth and continuous, with no rapid changes in direction or altitude, as would occur if the river contained sharp turns or encountered a waterfall. The molecules of water continue to move and shift positions gradually. Things are in an approximate state of equilibrium. Voters might be considered as analogous to these molecules, and the electoral system analogous to the river. Watching a smooth and slowly flowing river would be comparable to observing voters during especially calm electoral periods. Under such circumstances, voters might shift their positions on issues, and perhaps their partisan choices, but gradually and in small numbers at each election. The electoral system would appear calm just as the surface of the river would appear calm.

However, if the water in the river flows over a waterfall, a great deal changes quickly with regard to the relative positions of the molecules. The water appears churned. Yet, while the surface of the water may appear highly disturbed, there are patterns that can be discerned within the turmoil. Vortices and waves form at the lower end of the waterfall, eddies swirl, and currents develop that ultimately lead to a new positional balance among the molecules as they continue their journey down the river.

An electoral landslide can be considered analogous to a waterfall in the sense that extreme nonlinearities could be expected within the political fabric of the electorate during a landslide. Longitudinal nonlinearities require change, and in a landslide there is a great deal of change in the voting patterns of many citizens. But fundamental questions remain as to what happens when that change occurs, and this is where the usefulness of the river analogy ends. Do a certain percentage of voters uniformly distributed throughout large regions of the nation simply switch parties, or are the mass dynamics much more complex? Do some groups of voters simply stop voting while others surge to the polling booths in record numbers? Is there such a thing as an equilibrium state in the aggregate dynamics of a landslide?

Indeed, it is necessary to address the basic question of how an equilibrium is defined within the dynamics of a landslide. What does it mean to say that a society voted in a state of aggregate equilibrium? When little is changing, the concept of an aggregate equilibrium is easy to conceptualize as the relatively constant vote proportions. But when things are shifting quickly, how does one know if the political balance settled down to an equilibrium on election day? Does the nation arrive at a new equilibrium with one smooth transition? Does the transition of support from one party to another in various areas of the nation follow a nonlinear pattern to a new equilibrium? Or does the election simply bisect a rapidly changing dynamic political process that has not yet arrived at equilibrium? Indeed, do landslide elections really measure the nation's political sensibilities in a state of equilibrium at all? All of this is addressed below.

The question of whether an election measures a society in a state of political equilibrium is especially important here. Common wisdom tells us that elections, in general, measure a nation's political choices while in state of equilibrium. Recall the common phrase, "The people have spoken." The implication is that some firm decision has been made in an election. But it is odd that polls regularly track the trends in voter support during the election, often noting great potential for volatility and rapid change right up to election day. Why *must* election day be fundamentally different from the previous days? Indeed, it may be that its only special character results from the arbitrary coincidence found in the electoral calendar. To offer one example, as the 1976 election approached, many polls revealed that Gerald Ford was gaining popular support. But if the 1976 election had been held two weeks later, would Gerald Ford have defeated Jimmy Carter? What then does an election measure, a nation in equilibrium or a populace in motion?

In a landslide election, these questions become increasingly relevant. During tranquil political times, one could argue that the partisan balance changes little from election to election, and the elections probably reflect the nation's mood in an approximate state of equilibrium. But this cannot be so easily posited during a landslide. If the voter movements are sufficiently large, or the political setting substantially disturbed, it may be that the election simply measures the voters' mood at a point in time. The winner might not be different were the election held a month later, but there would be no guarantee that the partisan totals would be the same. In short, the nation may not be in a state of equilibrium. There is currently no evidence reported in the extant empirical literature on voting that conclusively answers these questions with regard to conditions of large magnitude electoral change. The current analysis takes aim at them directly.

The analysis begins with a discussion of some existing concepts of rapid electoral change. The focus here is on those studies that explicitly identify structures of aggregate electoral change, with particular emphasis on structures that allow for potentially explosive voter movements. A formal model of electoral change is then proposed and explored with regard to a complete collection of county-level aggregate data. Finally, an examination is made of some survey data collected in 1964 to allow for a general discussion of some of the issues that drove the voter movements in that crucial election.

The 1964 Election

To help put the current investigation into historical perspective, table 4.1 contains the level of partisan mobilization for the United States for all presidential elections in this century, from 1900 to 1988. All figures in the table are written as proportions of the total eligible electorate as determined by age (as well as gender before 1920). Vote mobilization measures are used throughout this analysis.

In addition to the mobilization figures for the Democratic and Republican parties, table 4.1 contains the difference in mobilization between each party, a baseline estimate of the level of new voter activity (measured as the difference in total mobilization between each election and the previous presidential

TABLE 4.1. Presidential Mobilization as Proportions of Total Eligibles, 1900–1988

Year	Democratic	Republican	Difference	New Voters[a]	Total
1900	0.302	0.343	-0.041	-0.055	0.665
1904	0.218	0.327	-0.109	-0.086	0.579
1908	0.250	0.299	-0.049	0.000	0.579
1912	0.227	0.126	0.101	-0.037	0.542
1916	0.309	0.289	0.020	0.086	0.628
1920	0.149	0.264	-0.115	0.279	0.437
1924	0.127	0.238	-0.111	0.004	0.441
1928	0.212	0.303	-0.090	0.079	0.521
1932	0.302	0.209	0.094	0.006	0.527
1936	0.348	0.209	0.139	0.045	0.572
1940	0.323	0.264	0.059	0.019	0.591
1944	0.285	0.245	0.040	-0.058	0.533
1948	0.254	0.231	0.023	-0.021	0.512
1952	0.272	0.338	-0.066	0.101	0.613
1956	0.249	0.341	-0.092	-0.019	0.594
1960	0.314	0.313	0.001	0.039	0.632
1964	0.379	0.239	0.140	-0.012	0.621
1968	0.260	0.265	-0.004	-0.011	0.610
1972	0.207	0.335	-0.128	0.038	0.552
1976	0.268	0.257	0.011	-0.016	0.536
1980	0.216	0.268	-0.051	-0.009	0.527
1984	0.215	0.311	-0.096	0.002	0.530
1988	0.229	0.268	-0.039	-0.028	0.502

[a]The new voter proportion for 1920 is the change in total voters between 1916 and 1920 divided by the number of eligible voters in 1916. This controls for the expansion in the number of eligible voters in 1920 due to the extension of the franchise to women. Similarly, the new voter proportion for 1972 is the change in total voters between 1968 and 1972 divided by the number of eligible voters in 1968. This is to control for the extension of the franchise to people 18 to 20 years old.

election), as well as total mobilization for each election. The new voter numbers are included to indicate easily which elections were accompanied by large increases in total voter turnout.

Note that the 1964 election contained the largest difference in partisan turnout among all of the elections. It is this observation that is used here to characterize the 1964 election as the "largest" landslide in this century.[3] In that election, there was a 14 percent difference in mobilization between the Democratic and the Republican parties. The second largest difference was Roosevelt's landslide win over Landon in 1936. The third largest landslide took place in 1972 when Nixon defeated McGovern. Interestingly, in each of these cases, the domestic setting in the United States was one of significant turmoil. The depression still gripped the nation in the mid-1930s, the 1960s was a period of intense civil rights activity, and the Vietnam War was shaking the social fabric of this nation in the late 1960s and early 1970s. While the reasons may remain unclear, these observations suggest that the simultaneous occurrence of intense domestic turmoil with each of the three largest electoral landslides in this century is not simply a coincidence. Thus, it is not routine politics that drive these elections.

Note, however, that of each of the landslide elections, the 1964 election contained the smallest change in voter turnout, as is evidenced by its relatively small-magnitude new voter estimate. While ecological considerations caution against making firm conclusions with these highly aggregated numbers, it nonetheless appears, on the surface, that some landslides contain substantially more new-voter activity than others, and a simple generalization across all landslides in this regard (i.e., whether they are predominantly partisan-switching or new-voter phenomena) may not be accurate. Nonetheless, other generalizations may yet emerge.

The Specification of Rapid Change

The model developed here is a portrayal of partisan competition. It isomorphically parallels in its algebraic structure many of the potential internal characteristics of an electorate experiencing rapid and large-scale changes in aggregate partisan support. In this sense, the model is a formal representation of the landslide phenomenon. The model is a system of two time-dependent and interconnected differential equations. The two equations model change in the aggregate support of the Democratic and Republican parties, respectively, between two elections. These types of formal models of social systems have been productively exploited in the social scientific literature by Coleman (1964, 1981), Simon (1957), Przeworski and Soares (1971), Przeworski and Sprague (1986), Sprague (1981), Tuma and Hannan (1984), Huckfeldt (1983), Huckfeldt, Kohfeld, and Likens (1982), Huckfeldt and Kohfeld (1989), Brown (1987, 1988, 1991), Gillespie et al. (1977), Ward (1984), and others. Outside of the social sciences, these models have found great use in the field of population biology (in particular, see May 1974). Moreover, the mathematical theory of all such systems is complete and well understood. This is true for both linear and nonlinear systems (see Hirsch and Smale 1974; Mesterton-Gibbons 1989; Kocak 1989; Luenberger 1979).

I begin the development of the model of partisan competition by focusing on three distinct mechanisms by which former supporters of one party would switch to support the other party. Electoral conversion is an important factor of such change that has been raised repeatedly in the realignment literature (Key 1955; Sundquist 1983; Erikson and Tedin 1981; Burnham 1970; Ladd and Hadley 1978). The first of these mechanisms, here referenced as the "dominance factor," captures the concept of voter movement being encouraged by the relative dominance of one party over the other. Algebraically, this is an idea that was first specified formally by Leslie (1948), and it is expressed by writing the equation for change in a party's support as a linear function of the ratio of that party's popular support to that of the other party. Substantively, the idea addresses voters' sensitivity to changes in a party's dominance within the local political milieu. In one scenario, such voters would be more inclined to switch parties as they find themselves increasingly outnumbered by others with opposing points of view. However, in certain situations, the reverse could occur, as when political minorities act to isolate themselves from (thereby resisting the pull of) political majorities. Substantively, both such mechanisms have been examined recently by Huckfeldt and Sprague (1987, 1988), and they are closely related to the "breakage effect" examined much earlier by Berelson, Lazarsfeld, and McPhee (1954, 98–101). Moreover, the sociological literature contains repeated reports of such milieu-dominating factors as influencing individual attitudes and behaviors toward conformity with the norms of the broader social environment (see Blau 1977; Blum 1985; Simmel 1955).

Thus, beginning with modeling change in support for the Democratic party, we have

$$dD/dt = q(D/R), \tag{4.1}$$

where dD/dt is the derivative that specifies change in the Democratic party between two elections, D is the proportion of the eligible electorate that supports the Democratic party, R is the proportion of the electorate that supports the Republican party, and q is a constant and a parameter of the model. Note that, with this specification, change in support for the Democratic party will increase as the local dominance of the Democratic party over the Republican party increases, *if the parameter* q *is positive.*

What is implied if parameter q is negative when estimated? It is important to understand that other terms are being introduced into the model immediately below that may dominate the recruitment process for the Democratic party. If the parameter q turns out to be negative, then the process of Democratic recruitment will be dominated by these other factors, and the negative relative dominance factor will reflect the process in which political minorities will act to resist the "pull" of the increasing political majorities. The micromechanism by which this will occur has been explored by Huckfeldt and Sprague (1987, 1988). While substantively the model can account for both positive and negative estimates for the influence of the relative dominance factor, my own expectation is that the other recruitment factors presented below will dominate the recruitment process and that q will be negative, reflecting the resistance of political minorities in lopsided political contexts.

The second mechanism discussed here by which former supporters of one party may switch to support the other party is called the "interactive factor." Conversions across parties can be initiated through interactions among voters. Democratic voters interacting both directly and indirectly with Republican voters can lead to political change among all voters. For example, under situations in which a Republican voter may be hesitant with regard to his or her vote choice, the behavior of Democratic supporters within the local environment can suggest alternative partisan cues that such a voter may follow. For this effect to be substantial, both parties need to be well represented in the environment (such as within competitive areas). Following the previous example, a large interactive effect would require a sufficient number of Republicans available for conversion as well as a sufficient number of Democrats available to offer alternative partisan cues.

This interactive factor is well represented in the political behavioral and the sociological literatures. Within-group conversations have been found to have a clear manipulatory influence on individual attitudes and behaviors (see Molotch and Boden 1985; Berelson, Lazarsfeld, and McPhee 1954; Huckfeldt and Sprague 1987, 1988; MacKuen and Brown 1987). Partisan cues may be transmitted indirectly as well, however. This has been evidenced repeatedly in terms of individual and group biases that result from perceptions of real-world facts (e.g., Garfinkel 1967; Gurwitsch 1962, 50–72). Moreover, on the aggregate level, such interactive influences have been found to be of substantial magnitude and crucial to the specification of large-scale voter movements (Beck 1974; Brown 1987, 1988, 1991).

Formally, the interactive factor can be included in the model by rewriting equation 4.1 as

$$dD/dt = q(D/R) + wDR. \tag{4.2}$$

The multiplicative term wDR captures the interactive influence between the parties. The phrase DR identifies the probability of interaction between the two partisan populations, whereas the parameter w characterizes the probability of conversion given the interaction.[4] Interaction terms of the type specified above are symmetric in the desirable sense of having the largest numerical value when both Democratic and Republican populations are large. Moreover, such interaction specifications are often encountered in the broad literature on contagion, communication, and diffusion modeling (Coleman 1964; McPhee 1963; Simon 1957; Przeworski and Soares 1971; Sprague 1976; Rapoport 1963, 1983; Huckfeldt 1983; Huckfeldt, Kohfeld, and Likens 1982; Huckfeldt and Kohfeld 1989; Brown 1987, 1988, and 1991).

The third mechanism of partisan conversion included here reflects the ability of a party's popular support to grow in simple proportion to its existing level of local popular support. In this sense, voters may be influenced by a party's campaign without regard to interactive or relative-dominance factors.

For example, if the Democratic campaign is effective, some voters may support the Democratic party in direct proportion to the level of success that this campaign is having locally. This is the simplest of the three mechanisms discussed above and is included formally in the specification

$$dD/dt = q(D/R) + wDR + uD, \tag{4.3}$$

where u is a constant parameter in the model and reflects the growth of Democratic popular support as a proportion of existing levels of Democratic support. The term uD specifies what is here labeled as the "proportional factor" of partisan change. By itself, the proportional factor expresses exponential growth or decay, classically defined.

In a landslide election, the above three factors may not capture the increased speed of partisan change that would be due to changes in the campaign's momentum. That is, it is possible that the entire rate of change in a party's support, as currently expressed in equation 4.3, may vary in an accelerated fashion as the party's support changes. This is the classic momentum concept, as people jump onto bandwagons, or jump off sinking ships. Thus, the influence of the parameters q, w, and u on the overall model may proportionately increase (or potentially decrease) as the level of D varies. However, it may be that the variation in influence may be based not only on the level of D but on the magnitude of D-squared as well. Squaring the level of Democratic support addresses the concept of momentum as a function of the "bunching" of Democrats. That is, as Democrats interact with other Democrats, as would be the case in situations with increasing numbers of Democratic supporters locally, then their own character as a persuasive group, with regard to other voters, changes. This is related to the idea that a mob is fundamentally different in character than a simple collection of individuals. As the individuals aggregate, the potential for explosive partisan growth accelerates. For this reason, both of the above influences on the model are referred to as "accelerator factors" and are included in the specification of the model as in equation 4.4.

$$dD/dt = (1 + jD + yD^2)(q(D/R) + wDR + uD). \tag{4.4}$$

In equation 4.4, j and y are constant parameters in the model. In the absence of parameters j and y, the 1 in the first set of parentheses allows for the specification of the model in its unaccelerated form, as in equation 4.3. The parameter j characterizes the acceleration of a campaign's momentum as proportional to the current level of Democratic support. The parameter y mediates the acceleration input that results from the level of D-squared, that is, the "bunching" influence.

Two additional inputs are needed in the model to complete the current specification of a landslide election. The first input is that associated with the mobilization of new voters. While table 4.1 suggests that new voters may not have played a great role in the 1964 election, their influence cannot, at this point, be ruled out (see especially arguments by Andersen 1979; Brown 1991; Converse 1975; Campbell, et al. 1960; Petrocik 1981). It may be that, in some areas, new voters entered the electorate in large numbers, whereas, in other areas, some groups of voters stopped voting. It is most likely that the influence of new voters will be greatest in areas with many potential new voters, that is, in areas with larger nonvoting populations. This input, proportional to the size of the nonvoting population, is added to the model in equation 4.5.

$$dD/dt = (1 + jD + yD^2)(q(D/R) + wDR + uD) + vN, \tag{4.5}$$

where N is the proportion of the eligible population that is not voting, and v is a constant parameter of the model.

The final ingredient to the model is the inclusion of the upper and lower limits to the growth and decay of Democratic popular support. Since D cannot decrease below zero, the model's behavior must be limited with that lower bound. Similarly, since the Democrats cannot possibly mobilize more than all of the eligible voters, the model must have an upper logistic limit of unity. These two limits are added to the now completed expression of change in Democratic popular support, as in equation 4.6.

$$dD/dt = [(1 + jD + yD^2)(q(D/R) + wDR + uD) + vN](1 - D)D. \tag{4.6}$$

In equation 4.6, all of equation 4.5 is captured in square brackets and multiplied by the expression $(1 - D)D$. The term $(1 - D)$ captures the upper bound of unity in typically logistic fashion, whereas the term D specifies the lower bound.[5]

The algebraic representation of change in Republican support is structured in a parallel fashion to that of Democratic change as expressed in equation 4.6. Thus, change in Republican support is written as

$$dR/dt = [(1 + pR + sR^2)(f(R/D) + aRD + eR) + gN](1 - R)R. \tag{4.7}$$

Here, p, s, f, a, e, and g are constant parameters in the model, and all have interpretations parallel to their Democratic counterparts as found in equation

4.6. The limits in growth and decay of Republican support are expressed by the multiplicative term $(1 - R)R$, again in a parallel fashion with regard to equation 4.6.

The complete model of partisan competition under the conditions of a landslide is the combination of equations 4.6 and 4.7. Equations 4.6 and 4.7 constitute an interdependent system of two differential equations. The system is nonlinear in both states (i.e., the variables D and R) and parameters. The system is entirely symmetrical between both parties. In total, the system is a fully bounded expression of change in Democratic and Republican support as structured by dominance, interactive, and proportional factors, all mediated by acceleration influences due to a campaign's momentum, additionally enhanced by the influence of new voters.[6]

Estimating the System

The data used in this analysis are the complete collection of county-level aggregate electoral returns for all of the approximately 3,000 counties in the United States and are for the years 1960 and 1964. The election returns have been combined with needed census material for all counties in order to obtain the number of eligible voters in each county as defined by age (21 years and older). All partisan data are expressed as proportions of the total number of eligible voters in each county.

Throughout this analysis, the aggregate data are broken down by southern and nonsouthern regions. By "southern" is meant the five Deep South states whose majority populations voted for Goldwater in 1964 (see Asher 1988, 30; Black and Black 1987). These states are Alabama, Georgia, Louisiana, Mississippi, and South Carolina. There are over four hundred counties in the Deep South. Thus, the analyses are conducted separately for the counties in the Deep South and the counties in the remaining states outside the Deep South. The results are then compared. The breakdown between Deep South and other areas also enhances the interpretability of the results, since the regions are divided clearly using the objective measure of whether a state was "won" by one party or the other. The breakdown itself is necessary, since, in 1964, there were actually two landslides. The first occurred in areas outside of the Deep South, in favor of Johnson, and the second occurred in the Deep South, in favor of Goldwater. A comparison of the two landslides offers an extraordinary chance to begin to discern, in more general terms, differences in the internal structures of landslides.

Estimating the model expressed as equations 4.6 and 4.7 is not trivial (see Judge et al. 1982, 633–63). As is characteristic of all such nonlinear systems of equations, it is only a stroke of luck if the equations can be uncoupled and linearized to allow the use of commonly available regression techniques, and this is usually limited to the simplest of such systems (see Tuma and Hannan 1984; Coleman 1981). In the above system, it is not possible to solve for D and R explicitly. Thus, it is necessary to leave the model in differential equation form and to obtain estimates of the system's parameters using a nonlinear least squares procedure. (See Brown 1991 for a detailed description of the estimation procedures.)

This investigation requires the use of aggregate-level data.[7] The available survey data for the period under study are not adequate, by themselves, to answer the questions posed here. Small sample sizes, insufficient variation within regions, and the absence of a panel-type longitudinal component are three reasons, but there are others. The model explored here is specifically written to address contextual interpretations of aggregate voter movements during an electoral landslide. Aggregate data have been used repeatedly in the extant electoral literature to address such matters, and the current analysis pursues a treatment, however sophisticated, of this same type of data.[8]

Results

The parameter estimates for the entire model are contained in tables 4.2 and 4.3. In each table, the estimated equations are separated by party. Table 4.2 contains the estimates for the two-equation model with respect to the data for all counties outside the Deep South. Table 4.3 presents the estimates for all counties in the Deep South. Each table also contains chi-square statistics that test the significance of each estimate against the null hypothesis that the parameter equals zero. This test is made with respect to each estimated parameter's impact on the model's prediction hypersurface (see Brown 1991). The Simon-effects estimate the relative impact of each parameter on the model with respect to the other estimates.[9]

In general, the model fits these data very well. With regard to the data for the nonsouthern counties, the model explains near or above 80 percent of the variance between the years 1960 and 1964. While the model does less well with respect to the data for the southern counties, the fit for change in southern Republican mobilization is nonetheless quite high. The fit with regard to Democratic change is lower, even though it is still substantial.

The lower fit for southern Democratic mobilization change reflects a less distinctly patterned regional Democratic behavioral response. Recall that the model characterizes patterned behavior among voter aggregates. When the aggregate behavior is not well patterned, thereby showing less systematic change, the model's fit is lower. This does not mean that the model is poorly specified with regard to landslide voting, for both of the nonsouthern fits and the southern Republican fit clearly suggest that the model characterizes the

TABLE 4.2. Parameter Estimates and Simon-Effects for Areas outside of the Deep South States

Parameter	Estimates	Chi-Square ($df = 1$)	Simon-Effects
Republican model			
p	0.4298	47998	0.00876
s	1.1671	35870	0.00751
f	0.2019	511656	0.02903
a	0.6374	86239	0.01197
e	-2.3071	28639381	0.20712
g	0.3898	320285	0.02332
Democratic model			
j	-0.6798	3640	0.00262
y	0.5727	615	0.00107
q	-0.4258	6870810	0.11130
w	0.7370	73554	0.01139
u	1.2843	2332876	0.06412
v	1.2536	4234042	0.08584
Goodness of fit			
Republican	0.794		
Democratic	0.861		

systemic longitudinal variation in these data quite well. But it does suggest that change in southern Democratic voting may be less well patterned than the change in the other cases, and this is a valuable insight.

Among all large groups of voters in the United States, few had a longer and more consistent history of support for a political party than white southern voters had for the Democratic party before 1964. While more about this is suggested later in this analysis, it is sufficient now to notice that heavily institutionalized voters (i.e., voters having long histories of consistently patterned electoral behavior) may be more resistant to large-scale changes in voting behavior than voters with lower levels of electoral institutionalization. In this sense, the lower fit for change in southern Democratic mobilization would likely reflect the model's ability to discern the partial breakdown of *some* of this institutionalized behavior, with further decay occurring after 1964, rather than the immediate collapse of most of it. This observation with regard to landslide voting corresponds empirically and in theory with observations made elsewhere by Przeworski (1975), Sprague (1981), and Brown (1991) with respect to the behavior of highly institutionalized voters. It also

TABLE 4.3. Parameter Estimates and Simon-Effects for Southern States

Parameter	Estimates	Chi-Square ($df = 1$)	Simon-Effects
Republican model			
p	0.7951	1866	0.00608
s	0.2159	6	0.00034
f	-0.2543	309069	0.07832
a	0.0318	1	0.00014
e	0.5039	9059	0.01356
g	1.7074	870747	0.13540
Democratic model			
j	-0.7163	217	0.00163
y	-0.4019	5	0.00024
q	-0.1153	16703	0.01511
w	0.0491	1	0.00015
u	0.2608	1389	0.00450
v	-0.3860	73969	0.03374
Goodness of fit			
Republican	0.627		
Democratic	0.206		

corresponds with arguments made by Black and Black (1987) with regard to longitudinal changes in the political behavior of white southerners since the 1960s.

Lessons drawn from the behavior of the estimated model are best obtained from the graphical analysis that follows. The graphical analysis is comprehensive in utility in drawing behavioral interpretations from the model. Before doing this, however, it is useful to note that all of the parameters used in the model played a useful role in characterizing these dynamics in areas outside the Deep South (as evidenced from table 4.2). However, for areas within the Deep South, some of the parameters are, from a statistical point of view, less well defined. This suggests that the change in southern voting was not as highly patterned as change elsewhere. This supports other research on both the micro and macro levels that suggests that highly institutionalized behavior is more difficult to change.

Among the principal partisan conversion factors contained in the model (i.e., the dominance, interactive, and proportional factors), the proportional influences clearly dominate with respect to nonsouthern Republican mobilization, as is evidenced by the relatively high magnitude estimate of the Simon-effect for parameter *e*. Substantively, this means that Republicans lost support more heavily to the Democrats in areas in which there were larger numbers of Republicans. Interaction with Democratic populations or the relative dominance of Republican strength had less of an impact. Interpreting from the aggregate level, this suggests that former Republican supporters abandoned the Republican party uniformly in areas outside of the Deep South. Undoubtedly, this occurred, in large part, because they did not like the message of the Republican campaign, not because the Democrats did so much locally to pressure them to change their partisan allegiance.

Among nonsouthern Democrats, however, the story is more complex. The negative estimate for the parameter q indicates that the Democrats actually lost some support in some counties in which Democratic dominance over the Republican party was greatest. Indeed, in a separate analysis (not shown here), it is found that a number of these counties are located in some of the peripheral southern states, but some such counties are geographically distributed elsewhere as well. As noted earlier, the negative estimate of parameter q indicates that, in this landslide election, political minorities tended to resist conversion to the Democratic party when the dominance of that party was greatest. This is an important point, since it suggests a confirmation of results using aggregate data that have recently been reported by Huckfeldt and Sprague (1988, 1987) using survey data. Also, note that Democratic support increased as a result of both interactive and proportional factors, as is evidenced by the positive estimates and substantial Simon-effects for the parameters w and u.

New voters also played their part in the 1964 landslide. Among the areas outside of the Deep South, new voters clearly helped the Democratic mobilization efforts, as is evidenced by the positive estimate for parameter v in combination with its substantial Simon-effect. In the Deep South, new voters played a dominant role in Republican mobilization. This can be seen from the positive estimate for parameter g together with its large Simon-effect as seen in table 4.3. Interestingly, the negative estimate for parameter v in table 4.3 suggests that the Democratic party experienced a degree of demobilization in some areas in the Deep South. Note also that the Simon-effect for this parameter is relatively large as well.

Again, the above discussion of the parameter estimates helps to introduce an interpretation of the behavior of the overall estimated system. However, a more complete picture can be obtained from a graphical analysis of the entire model.

Figure 4.1 is a phase portrait of the estimated system using parameter estimates for the nonsouthern region. Republican support is represented on the horizontal axis, and Democratic support is represented on the vertical axis. All support is measured as a proportion of the total eligibles. There are numerous trajectories represented in figure 4.1. Each trajectory is created

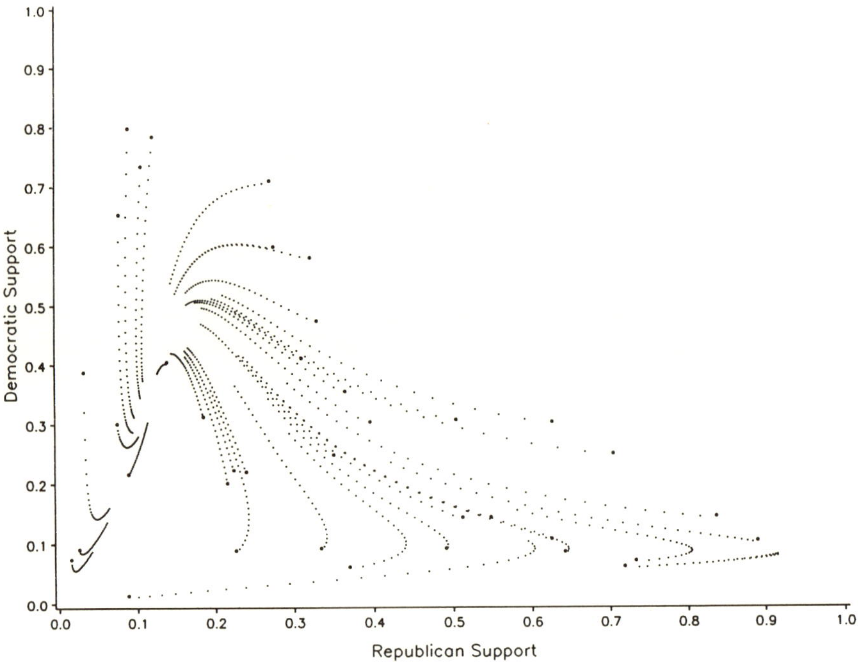

Fig. 4.1. Democratic and Republican partisan trade-offs using extended trajectories for areas outside of the Deep South

from the estimated system (i.e., both equations 4.6 and 4.7), using, for heuristic reasons, random initial conditions. Randomly chosen initial conditions help to demonstrate the great variety of behavior that is captured by the model.[10] The initial conditions are represented by the larger dots in the figure. The small-dotted trajectories that lead away from the larger dots represent the change in partisan support for each party as time progresses. Intuitively, the large-dotted initial conditions represent initial levels of support for the Republican and Democratic parties in 1960. In figure 4.1, the ends of the small-dotted trajectories do not represent the levels of partisan support in 1964. That is done in a later figure. In figure 4.1, the trajectories were allowed to continue far past their "natural pause" in 1964 in order to see more clearly where the trajectories were ultimately headed. The reason for this is explained more fully below.

It is best to begin the interpretation of figure 4.1 by noting some of its more general characteristics. Observe that the trajectories in the nonsouthern region follow a swirling pattern that somewhat resembles a vortex. Different things are happening at different places within the vortex. It is important to look at the major areas of change separately.

At low levels of initial Democratic support, there is often a considerable increase in Republican strength combined with little change in Democratic support. This indicates that Republicans gained in some areas outside of the Deep South in which there had been very little Democratic support in 1960. However, in areas with substantial levels of initial Democratic and Republican support, there is a large decrease in subsequent Republican strength combined with a comparable increase in Democratic support. This can be seen by following the diagonal trajectories up and to the left between the levels of .15 and .40 on the horizontal axis and .10 and .45 on the vertical axis. This is, surely, where most of the converting landslide in areas outside of the Deep South occurred. It happened in areas in which both parties had a substantial presence in 1960.

Interestingly, it is also clear from figure 4.1 that in areas with a very low initial Republican presence combined with very high Democratic support in 1960, there was a demobilization of some Democratic voters without a comparable increase in Republican support. In such areas, many Democratic voters simply stopped voting. At first glance, this may appear to be a surprising result, given the Democratic characteristic of the lopsided political victory. But, in 1964, there were some previously Democratic counties for which Johnson's new Democratic message was not well received. Yet in many such counties, the voters did not simply switch parties. Apparently, there was not a sufficiently large local Republican presence in these areas to enable these voters to complete the behavioral switch to support Goldwater. Some minimum threshold of Republican support was necessary in order to break, more completely, the institutionalized partisan bonds to the Democratic party. Indeed, as was mentioned earlier, a separate analysis of these data (not reported here) found many such counties in the peripheral southern region.

In general, figure 4.1 reveals a very complex setting of partisan change for the period beginning in 1960. Some areas demonstrated a Republican gain with little comparable Democratic gain; other areas demonstrated the reverse; whereas many other areas experienced a decline in Republican support combined with an increase in Democratic strength, the latter of which suggests evidence of a Republican to Democratic conversion process. Yet most trajectories seem eventually headed (following a number of turns along the way) for some equilibrium level of support of approximately .37 for the Democrats and perhaps a bit less than .20 for the Republicans. These numbers are, indeed, very close to the total levels of support obtained by both parties outside of the South in 1964. That the trajectories seem to be headed in the same direction as the overall means of support is a very good indication that the voting in areas outside the South did indeed achieve some level of regional equilibrium.

It is important to understand that, when aggregate voting is in equilibrium, it should not be expected that all counties have the exact same aggregate partisan balance. There will always be variations in partisan strengths across states and across counties within states. But when the aggregate trajectories point to an ultimate end near the actual vote proportions, it can be said that the movements of the trajectories are following the system's internal guidance around an equilibrium. In the language of dynamics, such an equilibrium is called a "stable attractor." Thus, we have begun to answer one of the questions posed earlier. The above interpretation of figure 4.1 suggests that landslides can result within an electorate that votes in a state of overall equilibrium. Later in this analysis, when examining the dynamic behavior of voters living in the Deep South, it will be clear that this is not always the case. But it is best, at this point, to complete the analysis of figure 4.1 by supplementing the trajectory information with that of a directional field chart. Such a chart is presented in figure 4.2.

In figure 4.2, the horizontal and vertical axes are identical to those of figure 4.1. The large dots evenly spaced in the figure are initial conditions similar to those in figure 4.1, with the exception that they are not randomly chosen. The lines that extend from the dots show the directions that any trajectory passing through that dot would take, which is why the plot is called a "directional field chart." The lengths of the lines reflect the speed with which the trajectory would travel at that point in the phase space. The small dots that have no lines coming from them are something else entirely and are explained below.

Note, in figure 4.2, that the directional field of the model for areas outside of the Deep South does, indeed, contain a topography similar to that of a vortex. While the concept of a directional field chart is not new (see Kocak 1989; Hirsch and Smale 1974), figure 4.2 is among the most complex directional field maps ever produced from estimates taken from actual data in the social sciences. It is very interesting to observe that social scientific data can indeed produce such topographies, and this, by itself, should encourage similar investigations of phenomena related to social and political systems. The results of figure 4.2 suggests that massive electoral movements of voters, both new and otherwise, need not be uniform in direction. Minimally, this is strong evidence to suggest that a landslide is much more than one candidate winning by a large margin. It is not just that one party's votes go up and the other party's votes go down. Patterns among contextually defined areas reveal complex forces of change.

One of the most interesting features of the directional field chart presented in figure 4.2 relates to the matter of a system equilibrium. The small dots in figure 4.2 (i.e., those that do not have directional lines emitting from

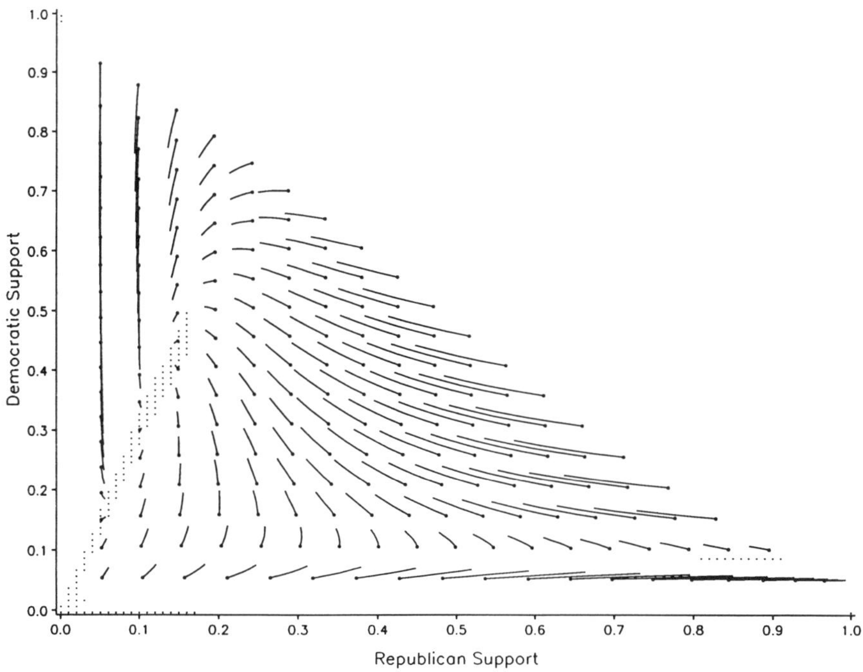

Fig. 4.2. Directional field chart for areas outside of the Deep South, 1960–64

them) produce a shading in the figure that identifies what is called an equilibrium marsh. An equilibrium marsh is different than an equilibrium point. Often an equilibrium point can be located within an equilibrium marsh, but this is not a requirement. An equilibrium marsh is an area in the phase plane in which the movement among all state variables becomes so slow as to nearly stop. By way of example, this could happen to a trajectory approaching an equilibrium point asymptotically, and very slowly. By the time the trajectory gets very far, the politics of the situation have changed. In our case, the election has come and gone. Thus, an equilibrium marsh is an area in the phase plane of Democratic and Republican competition in which change, for all intents and purposes, ceases. The mathematics of equilibria may indicate that further change toward a particular equilibrium point is possible, but the reality of the electoral calendar makes this observation irrelevant. Thus, the condition of voting in a state of aggregate equilibrium is defined here as follows:

Definition: A society votes in a state of aggregate equilibrium when both the final termination point of observation trajectories, as determined by the estimated system (i.e., when the trajectories are mathematically extended to the point of ultimate rest), and the actual aggregate vote totals exist within the phase area's equilibrium marsh.

At this point, it is useful to return to the portrayal of trajectories in figure 4.1. Note that many of the trajectories terminate in the area that is identified as an equilibrium marsh in figure 4.2. This indicates that the trajectories actually do, for all practical purposes, end in equilibrium, where an equilibrium is defined not in terms of a point but as an area of minimal change that contains within it an equilibrium point. This is an important observation because it leads us to ask how close trajectories of real counties get to the area of equilibrium. Recall that the trajectories of figure 4.1 are extended beyond that which would be typical for a particular county in order to see more clearly where the trajectories are heading. We now need a realistic portrayal of actual trajectories as bounded by partisan movements that occurred between 1960 and 1964. This is presented in figure 4.3.

In figure 4.3, the vertical and horizontal axes are identical with those of figures 4.1 and 4.2 with the exception that the ranges of the axes more closely correspond with the ranges of the actual data for Democratic and Republican support. The dots in the figure represent initial conditions for the system, randomly selected and typical of the real data. The trajectories that emit from these dots are the actual length of the trajectories as predicted by the model for the data for areas outside the Deep South.

Note, in figure 4.3, that the vortex pattern so clearly evident in figures 4.1 and 4.2 is still present, even if it is less distinctly visible due to the shorter length of the trajectories overall. Notice also that all of the patterns observed earlier with regard to the voter trade-offs between the two parties are still evident. However, the equilibrium characteristics observed in the earlier figures are not so apparent in figure 4.3 because the trajectories do not seem to get very close to the equilibrium marsh before the election occurs.

At first glance, it may seem that the election is cutting off the dynamic motion of voter support before equilibrium is achieved. Such an interpretation would lead us to believe that the electorate did not vote in a state of equilibrium and that the election itself merely took a measure of partisan support at an arbitrary point in time. But this is not what happened here. As mentioned earlier, it is unrealistic to think that all counties would ever have an equal, or even approximately equal, level of partisan balance. What is important, however, is that movement among so many counties does follow an identifiable pattern that contains an equilibrium marsh within which the actual mean vote proportions for the entire region are located. This is the requirement for a

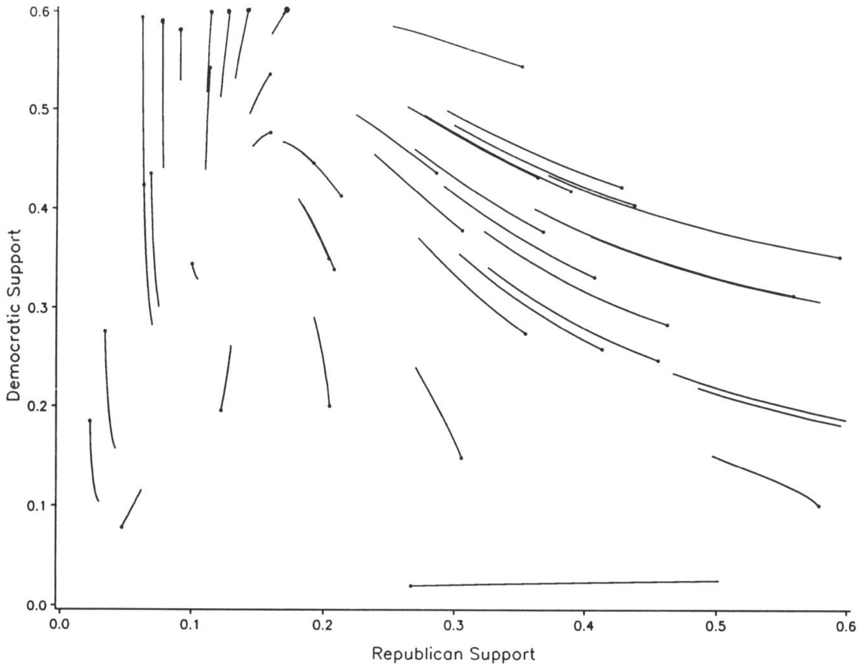

Fig. 4.3. Democratic and Republican partisan trade-offs using realistically bounded data ranges and trajectory lengths of areas outside of the Deep South, 1960–64

statement that suggests that an electorate's vote is a choice in equilibrium. It is not that all areas have equal partisan balances, but that the electorate, as a whole, has an identifiable center of balance, and that this center is very close to the actual outcome of the election. For even if, in an unrealistic and entirely hypothetical situation, the election were to be postponed until all areas had an equivalent partisan balance (i.e., until all trajectories ended within the equilibrium marsh), the outcome of the election would be no different.

But the landslide that occurred in areas outside the Deep South was much different than that which occurred in the Deep South. Figure 4.4 is a phase portrait for the Deep South showing extended trajectories. These trajectories are computed in exactly the same manner as was done for areas outside of the Deep South in figure 4.1.

Note some of the basic differences between figures 4.4 and 4.1. In figure 4.4, the pattern of aggregate electoral change does not "swirl" into the center of a vortex. Rather, there is a complicated collection of eddies, many of

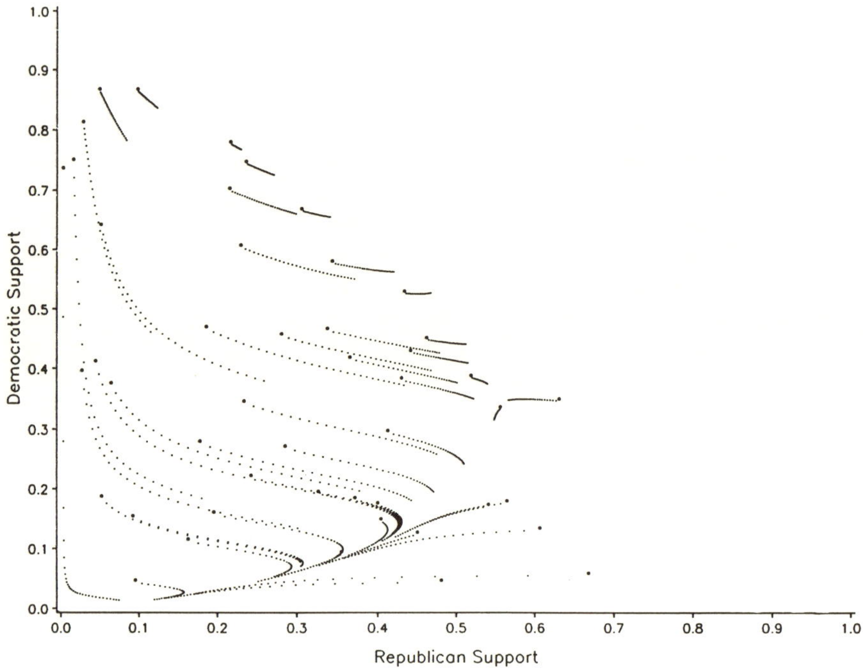

Fig. 4.4. Democratic and Republican partisan trade-offs using extended trajectories of areas in the Deep South

which seem ultimately to funnel down a diagonal path toward an equilibrium that has a very low level of Democratic support and a level of Republican support somewhat greater than .1. In addition, it seems that there may also be an equilibrium point near a Democratic level of support of .40 and a Republican level of support of .55 (discussed more fully below). Interestingly, some of the trajectories that begin with very low levels of Democratic support in the Deep South, but very high levels of Republican support, actually end up losing some of that Republican support without gaining Democratic support. There were very few such counties in the Deep South in 1960. But they did exist (particularly in the mountainous regions), and the model does accurately discern their abandonment of the party under Goldwater.

One of the most dominant characteristics of figure 4.4 is the loss of Democratic support in combination with a dramatic increase in Republican support. In particular, this occurred in areas with high or moderate initial levels of Democratic support combined with moderate levels of initial Republican support. That is where the largest elements of the conversion landslide occurred in the Deep South. Moreover, it is important to emphasize that there was *no* systematic movement in favor of increased support for the Democratic party anywhere in the Deep South. This is evidenced by the downward headings of all of the trajectories in figure 4.4, however wandering those downward paths may be.

Figure 4.5 helps guide our understanding of the location of the areas of partisan stability within the estimated system for the Deep South. Figure 4.5 is a directional field chart for the Deep South, identical in construction to figure 4.2, which was for areas outside the Deep South. As with figure 4.2, the equilibrium marsh area in figure 4.5 is represented by the dotted (i.e., shaded) area.

Note that the equilibrium marsh area in figure 4.5 has a curved shape and is quite extensive. The directional field pointers (i.e., the larger dots with lines extending from them) suggest that a stable equilibrium attractor does, indeed, exist at very low levels of Democratic support with moderate levels of Republican support. Another attractor (this time unstable, a virtual separatrix) exists at Democratic levels of support near .40 and Republican levels of

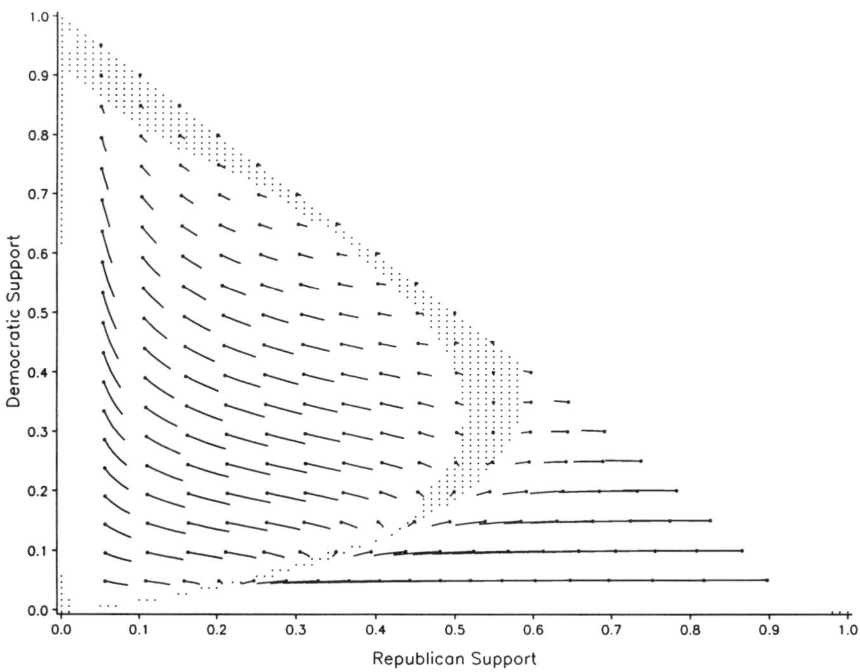

Fig. 4.5. Directional field chart for areas in the Deep South, 1960–64

support near .55, as suggested earlier with regard to figure 4.4. The appearance of this latter type of unstable attractor is not uncommon with models of this sort. Its appearance reflects the system's ability to discern faint subtleties in the dynamic characteristics of these data. But the substantive importance of such an attractor with regard to the stability characteristics of the overall system is dependent upon its proximity to the realistic ranges of the data. In this case (as demonstrated in the next figure), this attractor has a very minimal effect on the actual trajectories for counties in the Deep South and thus is of no consequence here.

The most interesting lesson to be drawn from figure 4.5 is that the actual proportions of support for the Democratic and Republican parties in 1964 are not contained within, nor are they near, the equilibrium marsh. For the Deep South, Democratic mobilization dropped from .17 in 1960 to .13 in 1964. Republican mobilization in the Deep South increased during that same period from .11 to .22. That these 1964 levels of partisan support do not correspond to areas near the system's equilibrium marsh indicates that the electorate in the Deep South was not voting in equilibrium in 1964. Indeed, this helps confirm our earlier suspicion with regard to partisan change and the Democratic party. In 1964, the Democratic party was a party in transition. A great deal of electoral institutionalization in favor of the Democratic party was eroded in 1964. But the entire Democratic house did not collapse in that election. In particular, certainly the departure of white southern supporters from the Democratic party continued well after 1964. The election itself, perhaps a watershed in directing the flow of the shifting partisan tides, was nonetheless only one step along a longer road of change. Indeed, this point confirms similar observations made elsewhere regarding the realignment character of the 1960s and 1970s in the Deep South (Black and Black 1987).

Where else did the Republicans gain the remainder of their support in 1964 if not entirely from the ranks of former Democrats? When discussing the parameter estimates in table 4.3, recall that Republican change between 1960 and 1964 in the Deep South was found to depend, in large measure, on the support of new voters. This observation was made with regard to the sign of the estimate for parameter g combined with its Simon-effect. In figure 4.6, a Republican dependence on new voters in the Deep South is clearly evidenced.

Figure 4.6 is a phase portrait for the estimated system with regard to the Deep South. The trajectories are *not* extended beyond the 1964 election. Moreover, ranges of the axes reflect the realistic ranges of the mobilization data for both parties. Figure 4.6 for the Deep South is comparable to figure 4.3 for other areas.

In figure 4.6, note that the dominant movement of the trajectories, however nonlinear, is down and to the right. This represents a decrease in Democratic support and an increase in Republican support. However, note that in

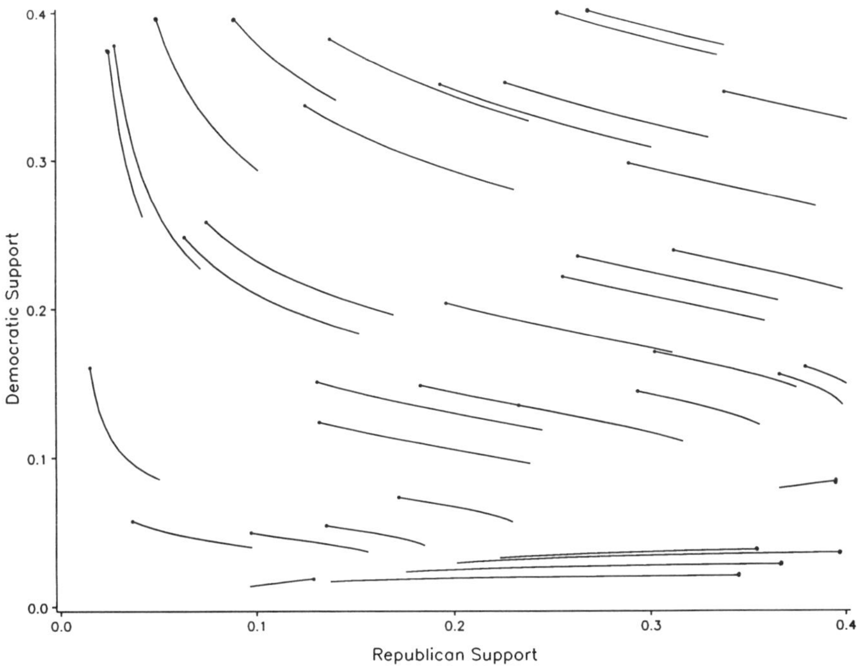

Fig. 4.6. Democratic and Republican partisan trade-offs using realistically bounded data ranges and trajectory lengths for areas in the Deep South, 1960–64

areas with more than minimal levels of initial Republican support in 1960, the rightward movement of the trajectories more than overshadows the downward movement. This suggests that, in such areas, the Republicans were gaining many more voters than the Democrats were losing. These voters were new voters, undoubtedly white, attracted by the new Republican message, and disenchanted by the relatively liberal Democratic campaign. However, note also, again in figure 4.6, that in areas with minimal levels of Republican support, Democratic support decreased without a substantial increase in Republican mobilization. In summary, the above observations clearly suggest that the 1964 election was characterized by a substantial demobilization of former Democratic supporters combined with some switching of voters from the Democratic party to the Republican party as well as a large turnout of new voters for the Republicans. The complexity and magnitude of these voter movements has not yet been thoroughly reported in the extant relevant electoral literature and, in general, is new to our historical knowledge of the Deep South.

A Note on Psychology

While the available survey data for 1964 are not adequate, by themselves, to answer the questions posed in the current investigation, they are nonetheless useful as a supplement to the aggregate analysis. In particular, survey data can remind us of the types of issues that drove that critical landslide.

Table 4.4 presents *t*-tests between means for various variables (mostly summary indices) contained in the American National Election Study of 1964 conducted by the Survey Research Center at the University of Michigan.[11] The means for each of the variables are computed separately for Johnson and Goldwater supporters. Sample sizes are not sufficiently large to break these means down by region, but the general psychological orientation of the campaign can still be discerned from the national means. The probability values in the table test the null hypothesis that there is no difference between the means. Thus, a small probability value suggests rejecting the null hypothesis.

The first variable listed in table 4.4 is the survey's general index characterizing each respondent's feelings about whether government is too strong. This variable addresses much of the tone of the campaign in 1964. Johnson was arguing for an increase in government involvement in local affairs, especially with regard to race and poverty matters. On the other hand, Goldwater was arguing strongly for the reverse. In the index, a score of one indicates that the respondent felt that government was too powerful, and a score of five

TABLE 4.4. Differences in Means between Johnson and Goldwater Supporters in 1964 for Selected Variables, Whites Only

Variable	Means Johnson	Means Goldwater	t	$p > \|T\|$
Federal government too powerful	3.77	1.77	17.87	.0001
Favors government aid to individuals	3.18	4.15	-8.99	.0001
Favors school integration	2.77	3.48	-5.74	.0001
Favors residential integration	3.55	3.33	1.80	.0721
Voting for or against a candidate	2.27	2.38	-0.82	.4125

Note: All *t*-tests are conducted under the assumption of unequal variances between groups.

indicates that the government was not too strong. From these results, it is obvious that this element of the campaign of 1964 was well discerned by the national electorate. Goldwater supporters clearly felt more strongly than Johnson supporters that the government was too powerful.

We can review the remaining results of table 4.4 in summary form with regard to the other variables. Johnson supporters tended to support governmental assistance to individuals whereas Goldwater supporters were generally more opposed to this. Johnson supporters tended to be more supportive of school integration than Goldwater supporters. There was not much of a difference, in general, between Johnson and Goldwater supporters with regard to whether residential housing should be integrated. This result suggests that educational integration was the more polarized, and thus more potent, of the two racially oriented issues. Finally, the supporters of both of the candidates tended not to differ with regard to whether they were voting for their candidate or against the other candidate. This final test is useful since the polarized nature of the 1964 campaign could suggest to some that voters were driven to the polls more by what they hated than by what they wanted. This could have happened to many voters. But these results suggest that it happened no more often to Johnson supporters than to Goldwater supporters.

It is difficult, using any data, to evaluate the voters' psychology so as to portray fully the intensity of feeling that many voters had during the 1964 election. But the above examination of survey data, however brief, does suggest that the differences between the two groups of voters were real, and the magnitude of the landslide suggests that the degree of intensity in these different feelings was quite high. Indeed, the intensity and tone of the political debate of that period are precisely the psychological factors that "triggered" the landslide to occur when it did. The aggregate analysis previously presented explains what happened to the voter aggregates, contextually defined, after the trigger was pulled.

Remarks

The single and most important result from these analyses is that aggregate voter movements within the context of largc-magnitudc and rapid clcctoral change can be extremely complex. Landslides are not simple matters of one candidate winning by a large margin. A significant rearrangement of much of the electoral landscape can occur in such elections.

In terms of the components of the estimated system investigated here, the results of this study suggest that the masses were, in part, guided in their partisan choices in the 1964 landslide election by the local dominance of one party relative to the other. They were also affected by interactive influences between supporters of the different parties as well as the simple proportional strengths of the parties. Moreover, acceleration factors related to the momentum of the campaign also influenced the mass dynamics of this landslide. These factors are contextual in structure, and significantly affect the speed by which the dominance, interactive, and proportional factors mentioned above act to mediate the aggregate voter shifts.

This analysis also finds that new voters played a substantial role in the 1964 landslide. In particular, new voters aided Republican mobilization efforts in the Deep South. However, again in the Deep South, the Democratic party experienced a substantial degree of demobilization that was not associated only with a Democratic to Republican conversion process. This occurred primarily in areas in which high levels of Democratic support in 1960 were accompanied with very low initial levels of Republican mobilization. This suggests, although it does not confirm, that a contextually conditioned threshold mechanism of partisan conversion operates that requires some initial level of opposition-party presence in order to initiate the conversion process. Interestingly, this idea finds an intellectual correspondence with arguments related to neighborhood change offered by Schelling (1978).

Landslide elections are not always elections in equilibrium. The analysis of the system's equilibrium behavior with regard to the 1964 election suggests that voting in areas outside the South was in equilibrium. The actual voting outcome for this area is contained within the estimated system's area of stability. But the 1964 election affected the Deep South differently. The Deep South was not in equilibrium when its voters went to the polls in 1964. The region was in the midst of a major electoral reorientation, only part of which was completed by the time the election took place. Both Democratic and Republican voting in the Deep South were just beginning to put on entirely new electoral faces. But the changes continued after 1964, and the 1964 election was merely one stop, however important, on a longer path of electoral evolution. This does not imply that voting in areas outside the South should have remained unchanged after 1964. Politics always change as societies evolve and face new challenges. But for areas outside the South, the 1964 election was a true point of rest, a moment of stability in partisan choice, an arrival at a collective equilibrium. Yet the evidence offered here suggests that the electorate in the Deep South had not resolved the electoral dilemma posed by Goldwater and Johnson in 1964 to the degree that occurred in the areas outside of the Deep South. Simply stated, when the election occurred, the voters in the Deep South ran out of time before they, as a region, were able to arrive at a new internal systemic balance.

The issues of the 1964 election drove this massive shift in partisan fortunes. The differences between the parties were real, and the voters could discern these differences. Indeed, these issues acted as the trigger for the electoral explosion that followed.

Again, one of the most interesting aspects of these findings is that the voter movements can be so violently complex, and so regionally varied on a systemic level once the shifting begins. This reinforces the concept that the social fabric of a polity is highly nonlinear, with each layer of society affecting other layers in an interdependent fashion. This addresses the complexity of the contextual nature of electorates, and reaffirms an understanding of politics as socially, not just individually, defined.

CHAPTER 5

Nonlinear Catastrophe Superstructures and the Fall of the Weimar Republic

The Weimar Republic collapsed, in part, as a consequence of Nazi intervention in the fragile and newly formed electoral system. But the experience of electoral collapse of such a system is not due only to the misguided motives of one group of people. There are always problematic groups in any and all electoral systems; but all systems do not fail. The problem of electoral collapse reaches back to the fundamental nature of the system's structure. It is not that the Nazis did bad things, but that the Weimar Republic was poorly grounded from the beginning in terms of longitudinal stability. This analysis investigates one aspect of that stability that has not been explored elsewhere. Indeed, this analysis explores a new and generalizable aspect of nonlinearity with regard to the larger context of political and social systems. It formally identifies large and highly nonlinear structures that control the more localized (i.e., small-scale) longitudinal change of critical system variables.

Most studies of nonlinear social systems focus on nonlinearity that appears within the range of the state variables. However, the nonlinearity that may appear within the range of such variables can be "controlled" by a nonlinear structure that resides mostly outside of this range. Indeed, looking at nonlinearity only from within the confines of particular variable ranges can potentially lead to a misspecification of the system if in fact the nonlinearity that does exist is a subset of a much larger nonlinear structure. Using a complete collection of aggregate electoral and census data based at the local level, this analysis discovers such a structure, here labeled a "nonlinear superstructure," with regard to the Weimar Republic. The analysis seeks to explain, at least partially, why the Weimar Republic was so easily destabilized by Nazi electoral participation. But the analysis also addresses the more broad theoretical issue regarding the potential and importance of discovering similar nonlinear superstructures elsewhere.

The Setting

There is a large literature concerned with German electoral behavior during the period of the Weimar Republic. Of course, most of this literature focuses on who voted for the Nazis during the later elections in this period. Good summaries of many of the debates in this literature can be found in Hamilton (1982), Brown (1982, 1987), and Lipset (1981). With regard to the current analysis, we need concern ourselves only with the historical basics.

There were eight elections during the period of the Weimar Republic. The dates of these elections are: (1) 6 June 1920, (2) 15 April 1924, (3) 11 July 1924, (4) 20 May 1928, (5) 14 September 1930, (6) 31 July 1932, (7) 11 November 1932, and (8) 5 March 1933. The NSDAP (i.e., the Nazi party), seriously contested all of the elections from 1928 on, however, their support in the 1928 election was minimal. After the 1928 election, the Nazi leadership decided to abandon their previous strategy of focusing on gaining support in the cities and rather attempt to recruit new voters in the rural areas. The party was very successful at doing this in the 1930 election. In the first election in 1932, three of the ideologically centrist non-Catholic parties that were newly formed during the Weimar period almost completely collapsed, with much of their support going to the NSDAP. (See Brown 1987 for a detailed description of this two-stage—new voter followed by voting shifting—realignment.)

Thus, growth in support for the NSDAP precipitated a rapid decline in the effective number of parties competing in the Weimar elections. This is another way of saying that the fragmentation of the Weimar electorate was reduced. Douglas Rae's fragmentation index is the most commonly used measure of the level of electoral fragmentation in an electorate (Rae 1967). This measure has also been used to measure market competitiveness in economic analyses (see Scherer 1980). Moreover, the measure has occasionally been adaptively modified in order to portray particular research needs (see Wildgen 1971, Taagepera and Shugart 1993, Molinar 1991, and Laakso and Taagepera 1979). A slightly modified version of Rae's original measure is used in this analysis. Originally, the measure is written as

$$F = 1 - \left(\sum_{i=1}^{n} T_i^2 \right),$$

where T is any party's decimal share of the vote and the sum is being taken across all parties for any particular election. The measure has a range of zero to unity, where a value of zero indicates a one-party system and a measure of one characterizes an extreme example of total fragmentation in which each person in the electorate has his or her own party.

It is important to note that Rae's fragmentation index is a probability measure. It measures the probability of partisan disagreement between any two voters chosen randomly from an electorate. The importance of this point to the current analysis is revealed in the model-building section of the analysis below. The modification to the fragmentation index that is used here is to base the measure on vote mobilization rather than vote share. Vote mobilization is derived from the pool of total eligibles rather than total vote. This is important here since nonvoters can be viewed as a separate group within an electorate. Previous analyses have indicated that new voters and demobilizing voters played crucial roles in the Weimar Republic (see Brown 1982, 1987). In this analysis, we are interested in changes in the fragmentation of the total electorate, not just among those who happened to vote in any one election. Therefore, the fragmentation index must include all potential voters as well as all voters.

Changes in partisan fragmentation occur when voters shift parties over time, demobilize entirely, or when new voters enter the electorate in proportions that differ from those of the previous participants. These voter changes typically affect the distribution of political power in the government. In this analysis, it is theorized that partisan fragmentation will change when the effective balance of political power changes. Effective political power is not related to vote mobilization as much as it is related to vote share, since it is the share of the vote going to a party that determines its ability to govern or to influence the government. Thus, if we are interested in investigating how partisan fragmentation changes in relation to changes in the electoral power of particular parties, we need a measure of deinstitutionalization that is based on vote share. In this analysis, Przeworski's measure for deinstitutionalization is used (Przeworski 1975). Here, deinstitutionalization is written as

$$D(t) = (1/2) \sum_{i=1}^{n} |\Delta V_i(t)|,$$

where $V_i(t)$ is the vote share of party i at time t. As with the measure for partisan fragmentation, $D(t)$ varies between zero and unity. At zero, the electorate is completely institutionalized, which means that there are no proportional changes in the participating partisan distribution of the electorate, and the distribution of effective political power can be assumed to remain relatively unchanged. At unity, the electorate is completely deinstitutionalized.

For any observed change of aggregate partisan vote share, the measure indicates the minimum proportion of the voting electorate that must have changed their support between two elections. Since change is counted twice after the summation sign in the measure (e.g., both from one party and to another), D is computed as one-half of the sum of all change across all parties. New voters influence the value of deinstitutionalization by directly acting on the level of V, causing one party to gain at the expense of other parties (something that cannot be gotten from a vote mobilization measure). Moreover, it does not matter if the number of parties changes between elections, since a disappearing party is simply given a value of zero for V at the appropriate time point. Thus, $D(t)$ represents the level of deinstitutionalization that an electorate experiences that directly affects the distribution of political power in the polity. It is this change in political power that causes governments to unravel and, as will be seen below in the case of Weimar, can lead to an electorate precipitously altering its degree of partisan fragmentation.[1]

Table 5.1 presents national averages during the period of the Weimar Republic for the three variables (1) support for the NSDAP as a proportion of total eligibles, (2) partisan fragmentation, and (3) deinstitutionalization. The variables, partisan fragmentation and deinstitutionalization, are defined as outlined above. Note that the 1928 support for the Nazis was minimal. The decision to move the campaign from the cities to the countryside in 1930 produced a remarkable increase in Nazi support. Moreover, the collapse of the ideologically centrist parties, combined with some new voter support more than doubled the 1930 support in the first election in 1932 (again, see Brown 1982 and 1987 for details regarding this realignment).[2] With regard to partisan fragmentation, note that the first election in 1932 experienced a large drop in the value of this variable. This corresponds to findings referenced

TABLE 5.1. Variable Averages, Weighted by Population

NSDAP Support as a Proportion of Eligibles	
1928	0.0194764
1930	0.1470843
1932A	0.3108654
1932B	0.2661057
1933	0.3908186
Partisan Fragmentation (Rae's index)	
1920	0.8242037
1924A	0.8534346
1924B	0.8359375
1928	0.8603006
1930	0.8519599
1932A	0.7849682
1932B	0.8191401
1933	0.7451895
Deinstitutionalization	
1924A	0.3151588
1924B	0.1724918
1928	0.2311663
1930	0.2755829
1932A	0.2303707
1932B	0.0724808
1933	0.1949013

above suggesting that the collapse of the non-Catholic ideologically centrist parties directly aided the NSDAP, thereby shifting three fragments of the electorate into one. Finally, note that approximately one-quarter (on average) of the electorate shifted their partisan support across parties throughout the Weimar years, as is reflected by the values of deinstitutionalization. This suggests high potential for a considerable degree of electoral volatility within that electorate.

This analysis focuses on the 1928–32 (July) period of the Weimar Republic. This is the period in which the Nazis gained their initial ascendancy, and this is also the period in which the electorate experienced its first and most important precipitous decline in partisan fragmentation. The data used here are at the level of the *Kreise*. These are local electoral and census geographical districts. There are approximately one thousand of these units in all of the Weimar Republic. The model investigated below is specifically designed to be evaluated with regard to such data, since the goal of this investigation is to explain change in aggregate partisan fragmentation with respect to aggregate changes in support for the Nazis and aggregate variations in deinstitutionalization. Thus, the focus of the study is to investigate the underlying structure of the macropolitical system. However, it is possible to include some microlevel (i.e., individual) reasoning when discussing macropolitical events. Indeed, the model developed below describes macrolevel change based on expectations of contextually conditioned voter behavior on the microlevel. (For a more complete discussion of the interrelationship between macrosystem models and microlevel behavior, see Brown 1991, chap. 3.)

The System

I begin the discussion of the specification of the system by focusing on change in deinstitutionalization. The theory advanced here posits that change in deinstitutionalization is expected to be influenced by both the isolated levels and the interactive levels of deinstitutionalization and partisan fragmentation. Thus, we write

$$dD/dt = bD + xF + aDF, \tag{5.1}$$

where D and F are as defined above, and b, x, and a are constant parameters in the model. In equation 5.1, the term bD describes exponential growth and decay in classic fashion, here identifying change in D to depend on the current levels of this variable. Also, future levels of vote shifting would be based on the current state of the electorate, and changes in the number of parties (or the proportional strength of the parties) would likely cause many voters to further evaluate their current partisan affiliations.

The interactive term, *aDF*, characterizes change in deinstitutionalization as dependent on the joint distribution of partisan fragmentation and deinstitutionalization. Thus, deinstitutionalization would be expected to change at a nonlinear rate, depending on the interactively combined values of *D* and *F*, rather than simply the isolated values of these variables. Substantively, this addresses the idea that the process of voter deinstitutionalization will be accelerated when both the current levels of fragmentation and deinstitutionalization are simultaneously high. This is a contextual expectation, and the literature on such contextual influences is both broad and conclusive (in particular, see Huckfeldt 1983; Huckfeldt and Sprague 1987, 1988, and 1992; Huckfeldt, Plutzer, and Sprague 1993; MacKuen and Brown 1987; Brown 1987, 1991). Voters react to their environment in ways in which one influence interactively compounds other influences. For example, one can easily imagine voters deciding to change parties when not only is their party losing terribly (perhaps to the point of extinction), but also many other neighboring voters are shifting their partisan affiliation. This is a "follow the crowd when your ship is sinking" expectation, and the nonlinearity of the interactive specification adds flexibility to the model's ability to characterize the contagion characteristic of such voter movements.

However, the Nazis played a special role in the Weimar Republic. Thus, there are two separate expectations with regard to equation 5.1. The first is that equation 5.1 as specified above will characterize many of the voter movements during the Weimar years. However, one would naturally expect that change in deinstitutionalization will be different depending on the level of local support for the NSDAP. Thus, voters may change their party affiliation depending on current levels of deinstitutionalization and partisan fragmentation. However, they may do so in substantially different ways when the new party in the neighborhood is as radically transformative as the Nazi movement. We can thus modify equation 5.1 to include this dual expectation by writing change in deinstitutionalization as

$$dD/dt = (bD + xF + aDF)(1 + eN), \tag{5.2}$$

where *N* represents popular support for the NSDAP as a proportion of the total eligibles, and *e* is a constant parameter in the model. Note that this specification for change in deinstitutionalization is both nonlinear and interactive with regard to expectations of Nazi influence on the other terms of the model. The 1 in the second pair of parentheses in equation 5.2 allows the model to estimate the influence of *D* and *F* on affecting change in deinstitutionalization independent of changes in Nazi support. The term *eN* allows for the interactive influence of local NSDAP activities in modifying these separate effects. This follows the dualistic substantive interpretation expressed above.

Change in partisan fragmentation is likely to be influenced by the level of deinstitutionalization, since fragmentation cannot change unless voters shift their partisan affiliations. However, we know from the historical record that partisan fragmentation plunged when the NSDAP attracted many of the voters who previously supported the non-Catholic ideologically centrist parties. Thus, we would expect partisan fragmentation to change depending on the current isolated level of deinstitutionalization as well as the interactive level of deinstitutionalization and support for the NSDAP. Following this reasoning, we can begin to characterize change in partisan fragmentation as

$$dF/dt = fD - qDN, \tag{5.3}$$

where f and q are parameters in the model, and the negative expectation of the term qDN reflects the depressing influence of the Nazi electoral movement on partisan fragmentation as expressed in the extant literature on this period.

However, it is useful to make a significant modification to the specification of partisan fragmentation as expressed in equation 5.3. Partisan fragmentation would also be expected to change based on its own current levels. However, the functional specification of this influence is not likely to be linear. The contextual literature suggests that voter movements are highly dependent on threshold limits on political behavior. Such findings date back to the "breakage effect" examined by Berelson, Lazarsfeld, and McPhee (1954, 98–101). The basic idea in the current context is that changes in fragmentation will be subject to a momentum effect (expressed as exponential growth or decay) due to contagion influences within the electorate. However, change beyond some level may level out (or even reverse direction) due to the dynamic process of voters reevaluating their participation in the current trends, perhaps as some voters rethink their initially emotional support for a party as radical as the NSDAP. In particular, these various changes can happen as the local political context changes due to increased support for the Nazis, resulting in more frequent interactions between NSDAP loyalists and everyone else.[3] Indeed, such a probabilistic interpretation of political change with regard to partisan fragmentation has a close correspondence with Rae's original definition of F. On the other hand, continued change in partisan fragmentation may cause a particularly rapid response from the electorate, one in which resistance to the Nazi movement collapses when many voters "throw in the towel," so to speak, and decide to give the Nazis a chance to influence change in the polity given that party's ability already to gain a significant level of popular support.

It is important to note that I am not claiming such multiple changes in the dynamics of partisan fragmentation *must* occur; I am merely stating that these changes are possible given the discussions that appear in the historical accounts of that period, and that the model should be able to reflect these changes should they exist. In summary, note that we are talking about three separate structural changes, however subtle, in the character of the dynamics of partisan fragmentation. First, initial exponential growth in fragmentation based on current levels of this variable, a modest decline in this growth rate as fragmentation approaches higher values, and finally a rapid change in fragmentation as the value of this variable reaches some breakage threshold value. This requires an algebraic specification on the order (at minimum) of a cubic with respect to F. We can now reexpress equation 5.3 as

$$dF/dt = fD - qDN + gF + jF^2 + mF^3, \tag{5.4}$$

where g, j, and m are parameters in the model. The relative signs and magnitudes of the estimated parameter values for g, j, and m will determine the structural topology of the attracting equilibrium surface for this model. In general, negative values for parameter g combined with positive values for either or both of the other two parameters indicate the appearance of variations on the back-and-forth dynamical theme described above.

It is possible to give greater probabilistic interpretation to the higher power terms of F, as mentioned above. As voters interact with other voters, they will see changes in partisan fragmentation, as is captured by the definition of F. But it is possible that voters may see such changes from multiple sources, not just one-on-one interactions. The squared and cubed terms of F capture this multiple conjoint influence in the sense of mutually corroborating observational interactions between voters and numerous contacts with their social environment. The higher the power, the greater number of sources of information exist from which voters are learning about their environment. More specifically, the higher powers of F are associated with identifying the changing influence of the collective millieu as fragmentation changes in a society. Intuitively, and as I have suggested elsewhere in this volume, this acknowledges the idea that a mob is different than a group of individuals, and it explicitly characterizes my own expectations that are associated with matters of contagion with regard to changes in this variable. Moreover, empirical testing as well as my own intuitive guidance suggests that a minimal specification of a cubic is appropriate for this initial identification of these dynamic processes.

Note that equation 5.4 now requires these structural changes in the dynamics of partisan fragmentation to be dependent on the scale of the variable F. There is no reason to assume that the scale of F, which is now arbitrarily set within the interval from zero to one, is the same as that which influences the micro–thought processes and macrodynamics that underlie the above defined change within the electorate. (Note that a primary inflection point with regard to F is fixed at $F = 0$.) Indeed, voter calculations, rational or otherwise, can be quite complex and subtle with respect to the values of social cleavage and breakage points. The theory outlined above identifies inflection points in the structure of change with respect to partisan fragmentation, but the electorate itself must tell us where those inflection points are placed with respect to F. Thus, to further generalize the model, thereby allowing the electoral data to characterize the scale and placement of the dynamic structural changes of the model with respect to F, we can allow the value of F to be scaled around a fixed reference point. The model now can be written as

$$dF/dt = fD - qDN + g[v(F - p)] + j[v(F - p)]^2 + m[v(F - p)]^3, \quad (5.5)$$

where v and p are constant parameters in the model. Here, p is the fixed value at which the central inflection point along F is placed, and v is the scaling parameter that identifies the magnitude of F that structures the character of longitudinal change in partisan fragmentation.

The final variable that requires specification is change in support for the Nazis. Beginning simply, change in support for the Nazi party is likely to depend on the current level of support for the party. This is captured as

$$dN/dt = cN, \quad (5.6)$$

where c is the exponential growth or decay parameter. However, change in support for the NSDAP is likely to be highly dependent on the values of deinstitutionalization and partisan fragmentation. In particular, it will be the combined values of these variables and support for the NSDAP that will likely structure the longitudinal dynamics of the party. This addresses an interactive specification with these variables that can be included in the model as

$$dN/dt = cN + kDN + yFN, \quad (5.7)$$

where k and y are constant parameters in the model.

Yet the historical literature suggests that heavy increases in support for the Nazis were followed by some aggregate declines in this support, again a function of the aggregate rethinking that was discussed above. This surge-and-decline concept requires a squared term with regard to N, since the decline will occur as a response to current supporters of the NSDAP interacting with other current supporters of the same party. This can be captured as

$$dN/dt = cN + kDN + yFN - wN^2, \quad (5.8)$$

Finally, increases in Nazi support need to be given an upward bound. The theoretical upper limit in Nazi strength is, of course, unity. However, it is not necessary that the realistic limit be the same as the theoretical limit. Indeed, we would like to know how close the realistic limit is to the theoretical limit.[4] We can add a scaling factor to the specification that allows for the identification of a more realistic upper limit to Nazi growth. Thus, we now have

$$dN/dt = (cN + kDN + yFN - wN^2)(1 - sN),$$

or, more economically,

$$dN/dt = (c + kD + yF - wN)N(1 - sN), \tag{5.9}$$

where s is the limit-scaling factor and a parameter in the model. There is no requirement that the parameter s be bounded between zero and one. If the theoretical and realistic limits are equal, then s would equal one. However, the estimated value of s would be greater than one if, as might be expected in certain contexts, the realistic limit to growth for the NSDAP were lower than unity. The estimated value of s would be less than one if the Nazi electoral movement was not near an aggregate equilibrium value when the election occurred. This latter case would indicate that the data contain little evidence of a "top-off" of Nazi growth at higher levels of aggregate support (i.e., the absence of a negative second derivative for change in N).

The description of the basic system is now complete. To summarize, it is comprised of the three equations, 5.2, 5.5, and 5.9, and is presented in collected form below.

$$dD/dt = (bD + xF + aDF)(1 + eN), \tag{5.2}$$

$$dF/dt = fD - qDN + g[v(F - p)] + j[v(F - p)]^2 + m[v(F - p)]^3, \tag{5.5}$$

$$dN/dt = (c + kD + yF - wN)N(1 - sN), \tag{5.9}$$

where

D = Deinstitutionalization (Przeworski's index)
F = Partisan fragmentation (modified Rae's index)
N = Mobilized support for the NSDAP.

One final modification to the system is needed. Previous research indicates that the Nazis had their greatest electoral impact in rural Protestant areas of the country (Brown 1982, 1987). Thus, the system is likely to display different properties when estimated for these regions as compared with those for the entire nation. Comparing both the national and Protestant rural settings also allows for a more detailed analysis of the overall content of the aggregate electoral behavior for that period. To accomplish this, it is necessary to estimate the system using both national and conditioned settings. This is done by first estimating the system's parameter values without conditioning, that is, estimating the system for the entire country. The conditioned system is then estimated by rewriting each of the parameters as a linear function of the conditioning variable. Using parameter s as an example, the new value of s would be, $s = s_0 + (s_1 COND)$, where s_0 is the national (unconditioned) estimate, s_1 is the conditioned estimate, and $COND$ is the conditioning variable. In this analysis, the conditioning variable is the interactive specification: the percent of the local population that is Protestant times an interval-level population-density measure (where one is rural and zero is urban at the extremes). This interactive specification is then standardized with a mean of zero and a standard deviation of one to yield a contextual measure that represents a Protestant rural milieu at the high end of the scale. A complete description of this conditioning procedure can be found in Brown 1991.

To summarize certain qualities of the system described above, the model developed and investigated here is a three-equation system of interconnected differential equations. The model is formal in the sense that it explicitly states in mathematical form the substantive theory upon which these discussions are based. The model is highly nonlinear, and the ability to evaluate such a high degree of nonlinearity with regard to large data sets is relatively new to social scientific research. Details of the estimation procedures used here can be found in Brown 1991. But it is important to note at this point that the estimation is achieved using numerically intensive techniques without uncoupling or linearizing the interdependent equations in the system. Similar methods used to estimate nonlinear dynamical systems have also been used and described with respect to social scientific applications by Przeworski and Sprague (1986) and Ward (1984).

The Results

The estimates of the entire system for both national and conditioned settings are presented in tables 5.2 and 5.3.[5] Table 5.2 contains the national (i.e., unconditioned) estimates broken down by the two time periods, 1928–30 and 1930–32 (July). Table 5.3 contains the conditioned estimates broken down similarly with respect to time. The measure of fit and the chi-square tests that are used in these tables are described thoroughly in Brown 1991. Briefly, the chi-square statistic tests the null hypothesis that the estimated parameter value

TABLE 5.2. Unconditioned Parameter Estimates

1928–30, Unconditioned			
Parameter	Estimates	Chi-Square (df = 2)	Simon F
f_0	0.14762	18386.27	0.08849
b_0	0.43679	720892.46	0.01640
m_0	0.19019	0.06	0.00006
a_0	0.44759	574893.44	0.01309
g_0	−0.42078	32159.06	0.16259
q_0	0.66407	1931.38	0.02144
s_0	0.99477	221800.15	0.09993
w_0	0.44192	5331.27	0.00674
v_0	0.34843	30791.99	0.15498
k_0	0.39001	8134.87	0.00056
j_0	0.12868	8.18	0.00041
y_0	0.85736	1248405.90	0.06521
c_0	1.15719	2320125.30	0.11994
x_0	0.20030	14953064.09	0.34984
e_0	0.47255	19427.89	0.00126
p_0	0.71071	504559.72	2.21682
Model Fits			
Partisan fragmentation	0.17648		
Deinstitutionalization	0.89172		
NSDAP	0.59111		
System average	0.55310		

1930–32, Unconditioned			
Parameter	Estimates	Chi-Square (df = 2)	Simon F
f_0	−0.50886	252323.97	0.13657
b_0	−0.02626	2482.18	0.00061
m_0	0.19554	0.03	0.00001
a_0	0.05023	5886.99	0.00052
g_0	−0.51405	40903.85	0.03870
q_0	1.20418	86676.55	0.06774
s_0	0.71597	450938.70	0.04322
w_0	0.25007	35121.89	0.00391
v_0	0.15501	42566.67	0.04007
k_0	0.38996	19852.91	0.00127
j_0	0.20683	14.68	0.00026
y_0	0.69248	2200365.86	0.14338
c_0	0.32736	804846.05	0.05165
x_0	0.27544	13608584.84	0.70822
e_0	0.15518	13883.93	0.00127
p_0	1.13798	494170.77	0.34764
Model Fits			
Partisan fragmentation	0.74820		
Deinstitutionalization	0.90017		
NSDAP	0.84543		
System average	0.83126		

TABLE 5.3. Conditioned Parameter Estimates

1928–30, Protestant Farm Conditioned			
Parameter	Estimates	Chi-Square (df = 2)	Simon F
f_1	−1.02490	27059.90	3.11438
b_1	0.04773	558.15	0.00224
m_1	0.48091	5220.36	0.03170
a_1	0.04569	480.07	0.00173
g_1	0.09054	23958.27	0.15920
q_1	0.81578	621.02	0.01295
s_1	0.83157	45927.49	0.03489
w_1	0.21620	113.70	0.00049
v_1	0.10649	188009.50	2.75878
k_1	0.01086	0.69	0.00002
j_1	0.20632	144.53	0.09134
y_1	0.23817	10050.84	0.00738
c_1	0.32051	15846.37	0.01585
x_1	0.00355	270.85	0.00193
e_1	−0.00456	0.44	0.00002
p_1	0.99656	28050.79	4.49959
Model Fits			
Partisan fragmentation	0.47100		
Deinstitutionalization	0.90320		
NSDAP	0.66512		
System average	0.67977		

1930–32, Protestant Farm Conditioned			
Parameter	Estimates	Chi-Square (df = 2)	Simon F
f_1	0.10608	685.31	0.005379
b_1	0.00340	2.57	0.000006
m_1	0.49962	0.14	0.000005
a_1	0.01026	15.74	0.000027
g_1	−0.02477	1.89	0.000090
q_1	0.12824	214.73	0.001016
s_1	0.21493	11312.82	0.002959
w_1	0.09514	689.88	0.000288
v_1	−0.00479	1.54	0.000022
k_1	0.01336	2.88	0.000005
j_1	−0.49641	15.70	0.000099
y_1	0.14054	5062.49	0.004771
c_1	0.09755	3564.60	0.003105
x_1	0.04945	4554.40	0.018841
e_1	−0.02376	66.80	0.000065
p_1	−0.06923	7.92	0.001186
Model Fits			
Partisan fragmentation	0.78454		
Deinstitutionalization	0.92251		
NSDAP	0.86730		
System average	0.85812		

is different from zero. The "Simon F" statistic is the change in the average fit for the entire system (i.e., the "system average" in the tables) when the parameter value is set to zero as compared with its estimated optimal value. A low value for a Simon F indicates that the parameter contributes relatively little to the model when compared with the other estimates.

As with most nonlinear systems, and due entirely to the complexity of the potential behavior of such systems, a detailed discussion of the estimated parameter values is the least productive way to proceed. Graphical analyses, as presented below, are more helpful in analyzing such systems from a heuristic point of view. However, it is useful to make a few initial observations regarding tables 5.2 and 5.3.

Note that the model fits these data quite well for both periods, as can be seen using the "system average" measure of fit. With regard to specific equations, only the equation describing change in partisan fragmentation for the early period has a relatively lower fit. This is expected since partisan fragmentation did not change dramatically between 1928 and 1930. However, note that the fit of the total system is higher for the Protestant rural areas (i.e., the conditioned estimates in table 5.3). Indeed, the fit for partisan fragmentation in the early period in the conditioned environment is considerably more than double that of the early period in the unconditioned environment.

With regard to the system limits in growth for the NSDAP (parameter s), note that in the unconditioned setting in the early period, the limit to growth for the Nazi party is near unity. This suggests that little harm would have been done had the model been written using the theoretical limit in this case. However, the value for parameter s is somewhat lower for the later period. Substantively, this suggests that the Weimar electorate in the July 1932 election may not have been in equilibrium at the time of the election. Again, the reasoning is that the data would show some indication of a slowdown of growth for the party at the higher levels of Nazi support if the electorate was approaching equilibrium. Thus, the model could not detect this slow-down *within the theoretical range of Nazi support*. This does not mean that the model is incorrectly specified! Indeed, the model specification is designed to detect such subtleties in the numerical historical record. Moreover, the substantive interpretation that is drawn from this result corresponds closely to historical accounts of the Weimar period. Indeed, the fact that a second election was held in 1932 only months after the July election is a clear indication that the Weimar leadership considered the mood of the electorate still to be in a state of flux at that time, and that potentially significant changes in the results were possible, albeit not certain. The fact that the NSDAP lost some support in the later election of 1932 may only point to the high degree of nonlinearity and high-speed volatility that existed in that electorate at that time.

Note that the conditioned estimates for the parameter s are quite large. This clearly indicates that the limits to Nazi growth are much more discernable in the Protestant rural areas where the Nazis did so well as compared with the nation overall. This is precisely what would be expected in areas in which the limits to growth would have been more closely approached due to the high levels of NSDAP support. This indicates that the model does well in capturing much of the complex underlying dynamic structure of these data.

One final observation with regard to a particular parameter is of special interest to the current discussion. Note in tables 5.2 and 5.3 that the parameter m is significant only in the early period in Protestant rural areas, as is indicated by the chi-square statistic. This parameter identifies the cubic term in the partisan fragmentation equation (equation 5.5). As will be seen in the graphical analysis below, it is in these areas that the system's dynamics are particularly complex. Moreover, it is not coincidence that these areas were the areas that produced the highest level of aggregate change in the critical early support of the NSDAP.

Important characteristics of the overall behavior of the estimated system can be discerned systematically using the following graphical analysis of the system's equilibria for partisan fragmentation. When something is at or near a state of equilibrium, change occurs slowly at best. If we are interested in identifying the determinants of change in partisan fragmentation, it is crucial to know where the variable does not change. This identifies centers of gravity for the system. It also identifies areas of attraction among the variables. If the trajectories are the participants in a dance, then the areas of equilibrium identify the boundaries and key turning points of the dance movements.

It is important to note that arrival at equilibrium with regard to one variable does not imply a simultaneous arrival at equilibrium for another variable. Thus, deinstitutionalization and support for the NSDAP can continuously change in the analysis that follows. Given particular ranges of these other evolving variables, we are searching for values of partisan fragmentation in which this latter variable begins to stabilize near an equilibrium surface. Normally, this would imply that the variable no longer changes significantly near these values, given the specified values of the other two variables. If these equilibrium values for partisan fragmentation are contiguous and form a surface, then change in another variable, say support for the Nazi party, "drags" the value of partisan fragmentation along that surface. If the surface ends or folds for some reason, and movement of a trajectory cannot remain on the surface given continued change in support for the Nazis, there is a possibility of rapid catastrophic change in partisan fragmentation as the value of that variable seeks a new state of equilibrium, wherever it may be.

Figure 5.1 presents some equilibria surfaces for partisan fragmentation with respect to a realistic set of continuous ranges for deinstitutionalization

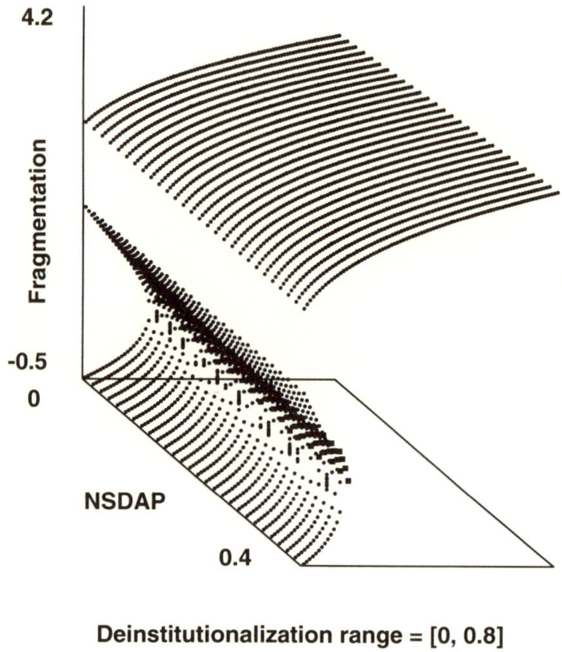

Fig. 5.1. Protestant farming areas, 1928–30

and support for the NSDAP. This figure is constructed for the early (1928–30) period for Protestant rural areas. Note that the vertical axis (partisan fragmentation) extends beyond the realistic range of the data. This is done to allow readers to see clearly the location of the surfaces, regardless of where they may be. There is a narrow band in the figure (in areas of low levels of deinstitutionalization) in which a curved equilibrium surface is more darkly shaded than in other areas. This band corresponds to a range on the vertical axis in which the values of partisan fragmentation fall within the actual range of the data on the vertical axis.

For areas within the realistic range of the vertical axis, there are no areas of equilibria for higher values of deinstitutionalization. In this case, the equilibria surface is far above the range of the data. Yet the existence of equilibria surfaces out of the range of the data is very important. These surfaces still influence the behavior of trajectories inside the area in which the data reside. This is an essential component of all dynamical systems. The behaviors of the systems are structured by the state space attractors. If one's data does not reside near an attractor, it does not mean that the attractor is inoperative. Indeed, the data—within its ranges—will be pulled in the direction of the attractor nonetheless. Thus, in order to understand the dynamics of such systems, it is crucial to know the location of the attractors, regardless of their location. To do otherwise leaves the myopic theorist without a complete explanation as to the reasons behind variable movements within more limited ranges. Indeed, figure 5.1 is an example of a picture that could lead to an incomplete understanding of the dynamics of the current model. What is needed is an expanded view—through a wide-angle lens, so to speak. This is done in figure 5.2.

Figure 5.2 is identical to figure 5.1 with the exception that the floor axes have been given expanded ranges beyond those of the data. It is in this figure that one can clearly see the shape of the single nonlinear equilibrium surface that is represented only as fragments in figure 5.1. This structure presented in figure 5.2 "passes through" the space occupied by the data of the state variables. Metaphorically, it is like a hand that dips into a large pool of water at certain spots, affecting the currents at those spots, but maintaining the existence of the remainder of the body away from those spots. This large surface is called a "nonlinear catastrophe superstructure," and its discovery (as estimated from actual historical data) is uncommon in the social sciences. It is

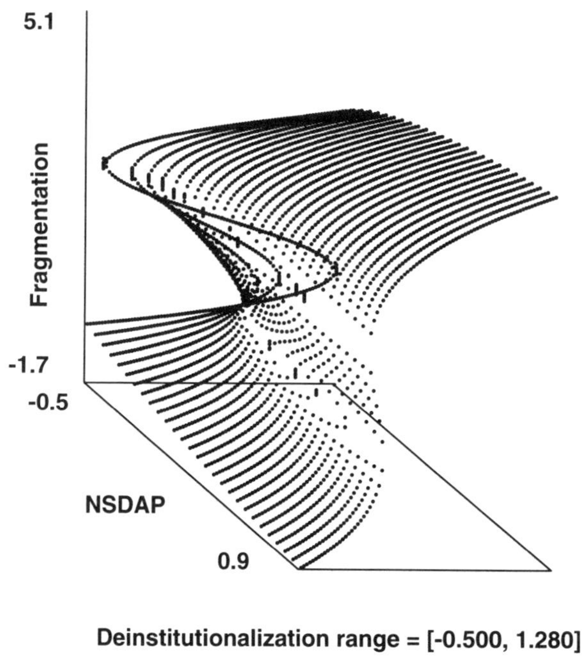

Fig. 5.2. Protestant farming areas, expanded axes

referenced here as a superstructure since much of its existence resides outside of the range of the data, while nonetheless influencing activity within the range of the data. It is identified as a catastrophe superstructure since there are areas on the surface in which continuous change in the values of the floor variables could cause rapid and sudden change on the vertical axis due to the inability of a trajectory to remain on the equilibrium surface continuously since the surface contains a variety of widely spaced folds. Such aspects of nonlinear dynamical systems are well known and are referenced in the literature of dynamics under the label "catastrophe theory." (In particular, see Thom 1975, Saunders 1980, and Abraham and Shaw 1992.)

Figure 5.3 is constructed to demonstrate how a catastrophe may operate within the current substantive setting, and as classically defined from a nonlinear system's point of view. The estimated system investigated in the current analysis is not used for figure 5.3, and thus the figure serves heuristic purposes only. Nonetheless, the procedures used to construct figure 5.3 are essentially the same as those used throughout these investigations for all of the other figures in which the actual estimated parameters are used.

In figure 5.3, there is a large backwards "S" that crosses most of the figure. That backward S is comparable to a cross section of the surface in figure 5.2 at some fixed level of support for the Nazi party. This backward S shape is the equilibria surface for partisan fragmentation, the variable on the vertical axis, given a continuous range of values of deinstitutionalization on the horizontal axis. In this figure, all trajectories of the state variables move in the direction of some point on this surface. The ultimate resting point on this surface depends on a variety of factors, one of which is the simultaneous arrival at an equilibrium value on the horizontal axis.

Many trajectories are also drawn on figure 5.3. These trajectories differ only in terms of their initial conditions. Given the parameter values used, note that most of the trajectories move in the direction of an equilibrium point on the left side of the figure. Note also that the movement of the trajectories is "guided" by the shape of the equilibria surface. There are two equilibria points on the surface that attract the trajectories. The equilibrium point at the lower right of the backward S-shaped surface attracts all of the trajectories that are to the right of this point. Of special interest are the trajectories that begin to the left of this lower-right equilibrium point. Note that these trajectories initially are attracted to the backward S-shaped surface. They then flow parallel to the surface in the direction of the left-most equilibrium point on the surface. When the trajectories arrive at the lower bend of the equilibria surface, they leave the surface since they cannot flow upward on the surface due to its movement in the direction of higher values of deinstitutionalization. The trajectories then move quickly to the uppermost equilibrium point on the surface. Similarly with regard to the results of comparable dynamics

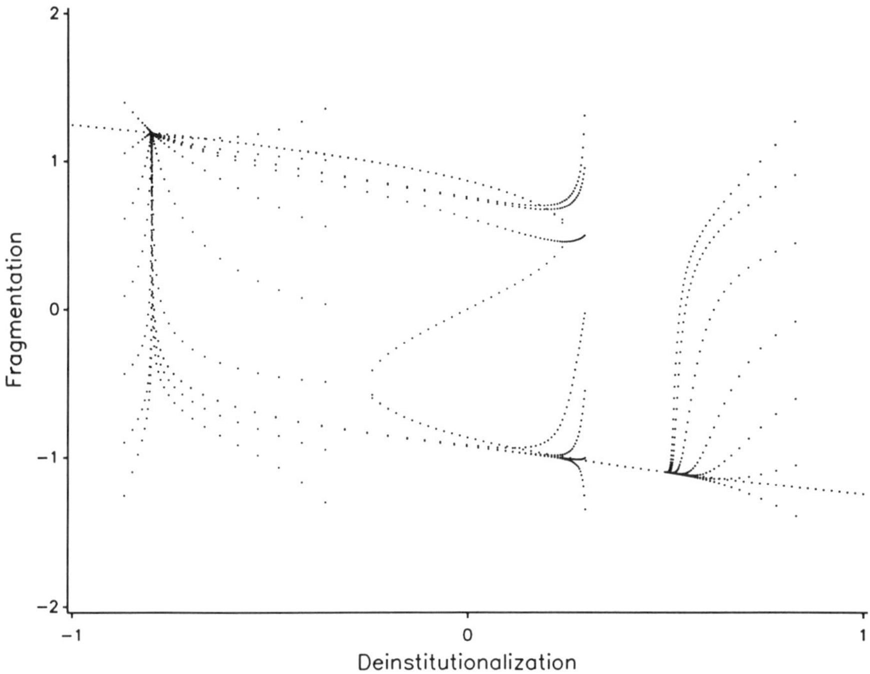

Fig. 5.3. Sample nonlinear catastrophe superstructure

investigated in chapter three of this volume, it is because of this rapid upward movement, combined with relatively small movement along the deinstitutionalization axis, that this type of phenomenon is referenced as a "catastrophe."

One of the reasons nonlinear catastrophe superstructures are so important is that the shape of the superstructures can change as the social and political context changes, thereby dramatically influencing the character of the variable trajectories. Indeed, the superstructures can sometimes have large areas located within the range of the data. In the absence of knowledge regarding such structures, it is possible to be unaware of the larger macrocontext of the social event. Indeed, it is possible that the entire nonlinear structure could collapse in size, placing much of itself within the range of data values.

Figure 5.4 presents a good example of how a large section of the superstructure can be found within the range of the data. This figure is constructed using the national (unconditioned) estimates for the early (1928–30) period, using floor axes that correspond to the actual ranges of the data values. In this situation, there is also only one continuous equilibria surface. However, due

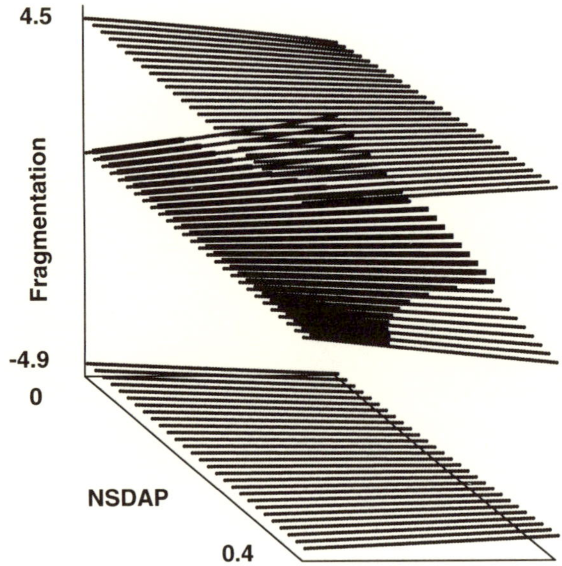

Deinstitutionalization range = [0, 0.8]

Fig. 5.4. National equilibria, 1928–30

to the limitations placed on the floor axes, it is possible to see only fragments of the surface, here represented as three nearly parallel planes. The darkly shaded area in the middle-level plane corresponds to a large area on the surface that has values of partisan fragmentation that fall within the actual range of the data. From figure 5.4 is it clear that partisan fragmentation was in a state of equilibrium throughout most of the country in this early period. Indeed, it is only in areas of very high levels of deinstitutionalization or very high levels of support for the NSDAP that the equilibrium surface left the range of the data for partisan fragmentation.

The results of figure 5.5 demonstrate how this situation can change rapidly. Figure 5.5 is constructed using the national estimates for the later (1930–32) period using floor axes that correspond to the actual ranges of the data values. Again, the darker shaded area of the surface corresponds to values of partisan fragmentation that fall within the actual range of the data on the vertical axis. Comparing figures 5.4 and 5.5 reveals how dramatically the equilibria surfaces can be changed. From figure 5.5, it is clear that partisan fragmentation was in equilibrium only in areas with very low levels of deinstitutionalization. In other areas, the trajectories were being pulled toward an

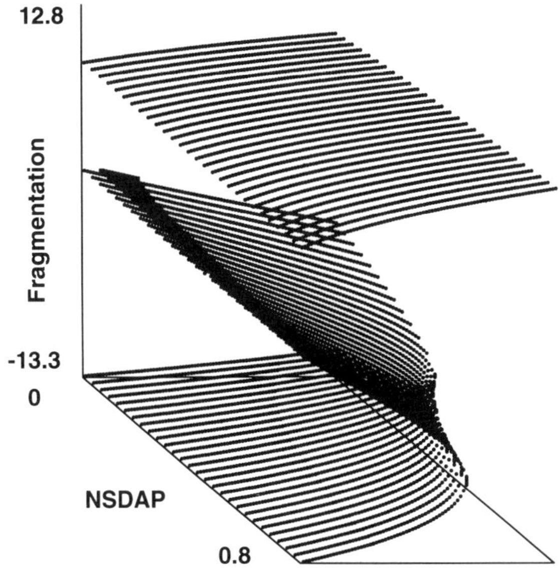

Deinstitutionalization range = [0, 0.7]

Fig. 5.5. National equilibria, 1930–32

"out of reach" equilibrium surface, either above or below, depending on the political context. The importance of this finding is to note that for much of the Weimar Republic, the country did not vote in a state of aggregate equilibrium in July of 1932 with regard to partisan fragmentation. This has theoretical implications with regard to the study of elections in general, and dramatically parallels findings reported elsewhere in which societies can experience electoral events in which an aggregate equilibrium is not achieved (Brown 1993).

Figure 5.6 is constructed using identical conditions as compared with figure 5.5, with the exception that the floor axes have been expanded in order to show a "wide-angle" view of the nonlinear equilibrium surface. The nonlinear catastrophe potential is again evident in this figure (due to the wide folds in the surface). Moreover, the superstructure characteristic of the surface is clearly apparent in this figure. However, note that much of the nonlinearity of the surface falls within the range of the actual data for the floor axes. Speculatively, the vertical position of the surface could easily shift in a different election, potentially plunging the areas of greatest nonlinearity and catastrophe potential into the entire range of data values. In a nonlinear dynamic world, keeping track of the influence of such equilibria surfaces on

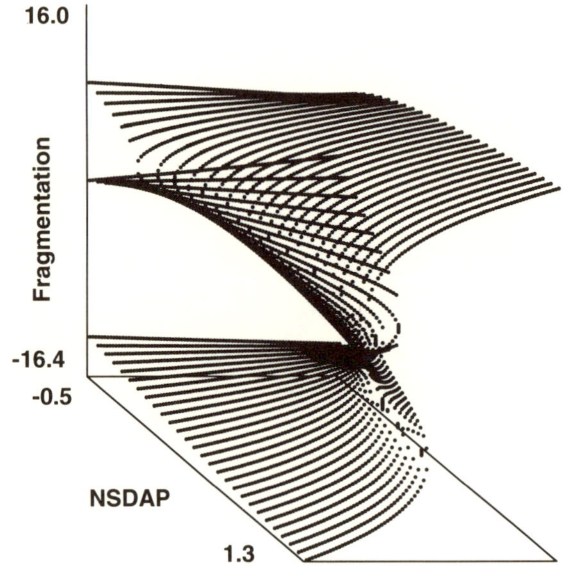

Deinstitutionalization range = [-0.500, 1.150]

Fig. 5.6. National equilibria, expanded axes

the structure of change in the state variables can be crucial to maintaining a holistic view of social theory.

One final observation regarding the total system behavior is useful at this point. It is not only the equilibria surfaces that change as the system transforms from context to context, and from time period to time period. Other aspects of the system evolve as well. Figure 5.7 displays one such aspect that can characterize nonlinear superstructures. Figure 5.7 contains a linearized stability analysis for the entire system in Protestant rural areas for the early (1928–30) period. Deinstitutionalization is held constant at the national average value for this figure. However, support for the NSDAP and partisan fragmentation are allowed to vary continuously using an expanded set of axes similar to those used in the earlier figures. Again, the expanded axes are used to allow for a wider view of the overall system behavior.

In figure 5.7, the maximum eigenvalue for the system's Jacobian matrix is plotted.[6] Higher values of the eigenvalue indicate rapid movement for the system's state variables overall, whereas lower values indicate slower rates of change. The surface is "localized" with respect to shades in order to show more clearly variations in the eigenvalue hypersurface. In this figure, note that

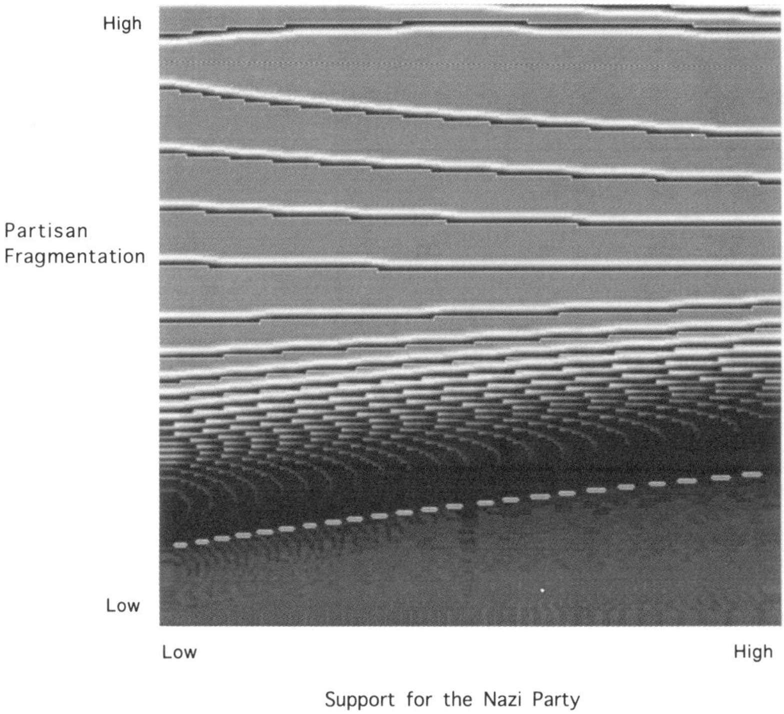

Fig. 5.7. Linear stability analysis of the entire system for Protestant rural areas during the 1928–30 period using expanded axes for the state variables, partisan fragmentation, and support for the NSDAP. Shades correspond to the maximum eigenvalue of the Jacobian matrix for the system given various values of the state variables. The level of deinstitutionalization is set at the national average.

various S shaped or curved patterns appear among the localized distribution of eigenvalues. This indicates that there is a large degree of highly nonlinear velocity variation in the dynamics of the system's trajectories. Substantively, in combination with the above results regarding the equilibria hypersurfaces, these results suggest that there exists a substantial potential for system volatility in these data. However, it is worth emphasizing again that this portrait of linearized stability presented in figure 5.7 will change dramatically when the political context, or the time period, is changed.

In summary, nonlinear superstructures have a variety of dynamic characteristics. These characteristics evolve in parallel with the evolution of society. The influence of these characteristics on the evolution of the state variables within their respective actual ranges is highly complex, and the study of this interrelationship is new to the social sciences in general. Within the context of the substance of this chapter, the mechanism by which this nonlinearity manifests itself within the Weimar electorate is behavioral contagion. It is not that the Protestant farmers migrated throughout the country after the 1930 election. Rather, the electoral dynamics that occurred between 1928 and 1930 in Protestant areas preceded and triggered similar mass behavior that subsequently developed elsewhere.

Conclusions

Nonlinearity is not just a local phenomenon. Indeed, the nonlinear physical world represents underlying dynamics that contain a logic that extends beyond that of manifested reality alone. This analysis examines one example in which such nonlinearity, here identified as a nonlinear catastrophe superstructure, is clearly identified from within the historical records of the Weimar Republic.

In the later years of the Weimar Republic, the Nazi party severely crippled the electoral system of that newly restructured democracy. The repercussions of that event led to tragedies on a planetary scale. Due to these repercussions, many people would like to have a relatively complete understanding of how it all began, perhaps to avoid similar events in the future in different temporal and geographic settings. That is, in fact, one motivation behind the choice of the historical subject in the current analysis.

Yet not all dramatic political events have such devastating human consequences. Nonetheless, many such events may reveal, upon analysis, a macrocontext within which the events took place. This investigation reveals a class of phenomena that could be quite general, and relevant, to many settings. Commonly, social scientists look at social phenomena within their precise settings, with variable values limited to historical realities. But, metaphorically, could it be that we have been focusing our attention too long on the puppet in the show, rather than standing back and observing the actions of the puppeteer as well? Is not the macrocontext within which the phenomenon resides equally important to examine, given the fact that whatever happens within the microsetting *must* obey the rules dictated by the larger context?

It is my perspective that the understanding of a social phenomenon's macrocontext is crucial to a complete understanding of the phenomenon. Indeed, this opens up a new area of social scientific investigation. Such macrocontexts are certainly not limited to the type investigated here. However, I suspect that the nonlinear characteristics of the macro superstructures described above may be a common quality of many such larger contexts.

This idea is actually a theme that has reappeared elsewhere (e.g., Brown 1993). With social phenomena, especially electoral phenomena, histories are punctuated by events that truncate and/or interrupt otherwise continuous processes. In the physical sciences, scientists can usually observe long histories of events, say, in the decay of a subatomic particle, or the orbit of a satellite. But imagine a situation in which a scientist observed a satellite for only one minute before it blew up. The satellite would still have had an orbit, even though the scientist could not observe its entire trajectory. The scientist would likely try to use the data for that one minute of observation to reconstruct the entire likely orbit of the satellite. That is the situation in which social scientists find themselves. A highly nonlinear social event is often abruptly terminated by, say, an election! If we confine our analyses to the variable movements that we actually observe, then we can never understand the larger picture within which the event took place. Moreover, we are likely to remain content with a false image of a linear world, since the nonlinearity that exists may be apparent only from a more distant theoretical perspective. Indeed, from the perspective of the above example, even an orbiting satellite's movement appears linear if the period of time within which the observations are made is sufficiently short.

Substantively, this analysis suggests that behavioral contagion is a powerful mechanism by which nonlinear dynamics can diffuse throughout a society. In the case of Weimar, something happened in Protestant areas that subsequently caught the attention of the electorate elsewhere. Evidence of the diffusion of nonlinear dynamical superstructures throughout a society in such a fashion is a phenomenon of mass behavior that yields new depth to our understanding of the nonlinear evolutionary potential of politics.

I am convinced, based on my own analyses and investigations, that human societies are as nonlinear as the remainder of the physical universe. But the way we observe ourselves may blind us to that nonlinearity. All human behavior takes place within a variety of contexts. One context is the nature of the localized milieu within which social interactions take place. Yet another context is the larger structure of the aggregate social relations. Examples of such larger structures are the nonlinear catastrophe superstructures that are examined here. Certainly there are other types as well. I have no idea, even speculative, as to the limits of the nonlinear nature of human life.

CHAPTER 6

Politics and the Environment: Nonlinear Instabilities Dominate

It is difficult to imagine many things that could be of as much concern to as many people as severe nonlinearities in the relationship between our forms of governance and the management of the planet's environment. Should there be, say, hidden and dangerous singularities or unknown oscillatory components to the dynamics of this relationship, our physical happiness, and perhaps our existence, could be threatened. In this chapter, I address such nonlinearities with regard to presidential politics in the United States. Nonlinearities are not limited to presidential politics and the environment, of course. But the current topic encourages a broader examination of related ideas in what is now emerging as a new subfield in political science.

The relationship between presidential elections in the United States and the degradation of the environment is not thoroughly understood. In this analysis, the relationship is characterized as a nonlinear interaction between oscillating environmental policy positions due to changes in partisan control of the White House and two other critical inputs. These inputs are public concern for the environment and the economic costs of environmental cleanup.

Broadly, this addresses a contextual question relating to political structure. The current theoretical connection between the dual cycles of partisan control of the White House and public concern for certain issues finds direct correspondence with recent research on cycles in the public mood by Stimson (1991). The joint effect of electoral structure and political context on this cycling is an extension of a thematic approach to the study of political parties as pursued by Sprague (1981), Huckfeldt (1983), and Beck (1974) in which party activities are seen as agents influenced by the structure of various social constraints. These constraints, combined with normal party activities, often produce nonobvious by-products, and that is what is being investigated here with regard to environmental policies. The results of the current investigation

Parts of this chapter are reprinted with permission from the article titled, "Politics and the Environment: Nonlinear Instabilities Dominate." *American Political Science Review* 88, no. 2 (June 1994): 292–303.

suggest that the interaction of the inputs mentioned above can lead to a surprisingly high potential for significant, and perhaps dangerous, volatility in the quality of the environment.

The relationship between democratic political structures and the environment is a relatively new area of research. It is easy to see how a rigidly communist or dictatorial society could inflict substantial environmental damage locally, or in some instances even globally, since public pressure for such governments to be environmentally conscientious is typically lacking.[1] But environmental damage occurs in democratic societies as well, and the level of this damage can be severe indeed, despite ostensibly "popular checks" against blatant abuse.

During the past 20 years, there has been an increasing global awareness of the immediacy of our planetary environmental worries. In part, this is evidenced by the existence of international meetings dealing with the environment, such as the June 1992 meeting in Rio de Janeiro of the United Nations Conference on Environment and Development. However, what has become increasingly obvious in international negotiations relating to the environment is that the dominant consideration of each of the participants is the likely impact on their own domestic politics. What is not well understood is how the organizational structure of each society's politics may influence longitudinal change in national environmental policies, and thus long-term change in environmental degradation. In short, understanding the domestic aspects of environmental policies goes well beyond obtaining recent survey data of the general public's attitudes. If serious damage is being done to the environment, understanding how governmental structure, particularly democratic political structure, contributes to the outcome is critical to our being able to minimize that damage in the future. The focus here is on an aspect of the electoral structure of presidential politics in the United States as an example of this relationship between democratic political structures and environmental damage.

Crucial to this discussion is the understanding that, in democratic societies in general—and certainly in the United States, environmentalism is subject to the same fate of political trade-offs as is any other concern. The typical environmental trade-off is economic, as measured in terms of losses in gross domestic product or jobs. Indeed, this is the focus of a recent analysis of the economic consequences of global warming by Schelling (1992). In one respect, the current investigation is an attempt to extend the discussions between environmentalists and economists to a broader range of social scientists as well, including political scientists who are interested in democratic electoral politics.

This analysis proceeds by developing a model of change in environmental degradation that is structured by electoral change, public mood, and economic cost constraints. Simulations are then performed on the model that reveal its more basic dynamic properties by systematically changing one input at a time while holding all others constant. An analysis of some of the global properties of the model is then conducted over a continuous and realistically wide range of potential parameter values important to the process. Substantive conclusions are then drawn from this characterization of the political-environmental process.

The Extension from Economics to Politics

Most of the social science literature on environmental degradation points to economic consequences of particular environmental change (e.g., global warming, ozone destruction, toxic contamination, etc.), including governmental responses to these changes. Particularly far-reaching in this regard is the analysis of global warming by Schelling mentioned above. Other prominent examples of the environmental-economic connection include Darmstadter (1991), Moulton and Richards (1990), Nordhaus (1991a, 1991b), and Poterba (1991).

On a more political level, one focus of the extant literature has been on the bureaucratic or regulatory responses to competing political and environmental demands. Tobin's analysis of the regulatory failure associated with biological diversity is seminal in this regard (Tobin 1990). Also, Wood (1988) and Wood and Waterman (1991) have produced two pioneering pieces of research that seek to identify some of the political determinants of longitudinal changes in environmental regulatory practices.

A difficulty in studying the interaction between politics and the environment is the current lack of data with regard to long-term environmental degradation. However, we are not at a total loss in this regard. Environmental modelers routinely develop models in the absence of data to explore the consequences of environmental changes to a broad range of interests, including ecosystem stability (May 1974), global warming (Energy Modeling Forum 1992), and other aspects of planetary transformation (see *New York Times*, 5 May 1992, B5–7[N]). In such cases, models are developed in anticipation of the data that are surely to arrive eventually.

In the current absence of data, model simulations are used to understand the environmental consequences of human action. Indeed, the current separate worries regarding ozone holes and global warming are a response to just such model simulations. In part, simulations are useful because a society may want to *avoid* a particular consequence of human activity. Thus, data for catastrophic scenarios can, hopefully, never be collected if the simulations lead to policy changes that avoid the catastrophes. In fact, simulations of plausible models are the *only* path available to us to understand the consequences of a given course of action without actually performing the action and waiting to see if disaster strikes in, say, one hundred years.

Yet models and simulations are built upon an understanding of the basic inputs of a process. Fortunately, we know a great deal about democratic electoral processes in the United States. Relevant to the current investigation, we know how constituent pressures are aligned differentially with regard to the political parties, and how governmental policies reflect these constituent pressures. Moreover, in the social sciences more broadly, there exists a long and rich tradition of exploiting models similar to the one developed here (with or without data) using simulations. This includes areas of research as diverse as political economy (Hibbs 1977), contextual theories of voter activity (Huckfeldt 1983; Przeworski and Soares 1971; Przeworski and Sprague 1986; Brown 1991), arms races (Richardson 1960; Ward 1984), and racial segregation (Schelling 1978). Thus, we have a methodological tradition as well as sufficient substantive knowledge to begin investigations into an area of research that we might call "political ecology."

The Model

In constructing a model of longitudinal change in environmental degradation that will be heuristically useful for analysis with simulations, it is wise to follow two basic guidelines.[2] The first is that the model should be general. Thus, model complexities should be held to a minimum so that it is relatively easy to identify change in the model behavior caused by varying each input. The second guideline is that it should be easy to identify the type of input each of the components of the model contributes to the model. In this case, the types of inputs will be limited in a classical fashion to gains and losses.

We begin by constructing a model describing change in environmental damage. Beginning with the gains (i.e., increases in degradation), a straightforward approach is to assume that environmental damage will increase logistically. At first, pollution will increase exponentially as industries grow and populations consume more products that harm the environment. However, this cannot continue forever, even in the complete absence of environmental legislation. In a worst case scenario, people would eventually die, perhaps of starvation if the polar ozone holes spread to temperate zones, leading to diminished agricultural production due to the higher levels of ultraviolet radiation. But in a less draconian fashion, one would expect that environmental damage has some upper limit beyond which a society will not willingly go. To ease the discussion of the model, the level of environmental damage, X, will be scaled to have an upper limit of unity. Thus, increases in environmental damage can be expressed as

$$dX/dt = rX(1 - X), \tag{6.1}$$

where r is a constant parameter of the model and represents what we can label as the "pollution growth rate" parameter.

Some readers may wonder whether it is correct to model environmental degradation as a smooth growth process since it may seem as if occasional spurts of activity would make the "ride" more bumpy. As with all models, the model developed here contains some simplifications, and a degree of "smoothing" is a desired trait of all attempts at segregating systematic and stochastic components of longitudinal change. However, many spurts in environmental degradation may be accounted for by the complicated oscillatory components of the model in its more fully developed form as developed below.

It is necessary to include two separate loss terms in equation 6.1. First, governmental environmental policy can act to limit or repair environmental degradation. This occurs, for example, when governments clean up toxic waste dumps, prohibit lead in gasoline, or perhaps in the future, if governments are required to orbit large mylar balloons to reflect some amount of solar radiation in the event of catastrophic global warming (see Schelling 1992). The amount of governmentally inspired reduction in environmental damage is likely to depend on the interaction between the amount of damage that exists, and the current level of public concern for that damage. This refers to a policy connection to longitudinal cycling in the public mood that can be deduced from empirical work by Stimson (1991), MacKuen (1981), and others, and has played a notable role in theoretical specifications involving cycling catastrophe models of environmental change by Rowland, Lee, and Goetze (1990). The theoretical justification for a specification for this loss is explained more thoroughly below.

The second loss term to include with equation 6.1 addresses the ability of the environment to repair itself over time. Pollutants tend to decay, be they chlorofluorocarbons or toxic wastes. Some have short half-lives while others stay around considerably longer, but they virtually all decay eventually.

Thus, we can now write the complete expression of change in environmental damage as

$$dX/dt = rX(1 - X) - pXY - kX. \tag{6.2}$$

Here, the variable Y is the current level of public concern for environmental damage, and the parameter p identifies the effectiveness of governmental policies in reducing current levels of damage as an interactive function of the level of damage and public concern for the environment. The parameter k is a decay parameter that reduces environmental damage based on some proportion of current levels of that damage. In this analysis, k is set equal to the expression, $-\ln(0.5)$/halflife, where "halflife" is the number of years before one-half of the damage decays by itself.[3] This allows us to examine the model based on how long a particular pollutant is expected to stay around once it is released into the environment. For immediate purposes, the term *halflife* can be thought of as an average half-life for current and total environmental pollutants. Note that the entire expression in equation 6.2 limits the level of environmental damage within the range zero to unity. This convenience further generalizes the model to an acceptable range suitable for numerical investigation.

The second major input into the environmental relationship identified above is public concern for the environment. Again, we will have gains and losses in public concern. As with environmental damage, it is reasonable to place a logistically defined upper limit on public concern. In this case, public concern for the environment will increase as damage to the environment increases up to some limit (call it the "panic" limit).

However, the public does not typically react to current levels of environmental damage in an instantaneous fashion. There is usually a lag in public concern as people wait until the environmental damage begins to affect them directly. Since the direct effects of environmental damage are usually due to pollutants introduced into the environment some time previously, the public is actually responding to an earlier level of environmental damage. Indeed, this lag can be as short as a few years, or as long as many decades, depending on the particular type of environmental damage being considered. For example, it took a few decades for environmental pressure to build substantially with regard to the poisoning of the Great Lakes, yet concern for the ozone holes and global warming increased more quickly once the connections between these phenomena and human behavior were identified. For modeling purposes, we can say that there will be an average lag in public concern. Thus, public concern will increase as a function of previous levels of environmental damage up to some limit. As with environmental change in equation 6.2, it is convenient to set this limit at unity.

Decreases in public concern for the environment are most likely to be due to the costs of cleanup. If people are going to have to pay substantially higher gasoline taxes, or taxes of another type, concern to clean up the environment is likely to diminish on average. Thus, the total expression for change in public concern for the environment can be written as

$$dY/dt = X_{old}(1 - Y) - Z. \qquad (6.3)$$

Here X_{old} is the lagged value of the level of environmental damage, and Z is the cost associated with cleaning up the environment. Growth in public concern for the environment continues as long as concern is not yet near its limit and costs are relatively low. Once concern is near its limit, or costs are high, costs will tend to dominate the dynamics in equation 6.3 and concern for the environment will begin to diminish.

The costs of cleaning up the environment will vary as well. There will be gain and loss characteristics of the change in costs. With regard to the gains, costs will tend to increase as *both* concern for the environment and actual environmental damage rise. Concern for the environment will spur politicians to address environmental issues, but politicians will allocate money for the environment only when there are clearly definable problems that can be addressed. Yet there are limits to the funds available for all governmental projects. For numerical purposes, we can scale the variable for costs such that the upper limit for governmental spending on the environment is unity. Thus, spending for the environment will increase in the direction of this limit as long as there is increasing public concern for the environment and sufficient current environmental damage.

Longitudinal decreases in governmental spending on the environment are likely to be due to the magnitude of the current burden of environmental spending. Environmental spending will tend to increase as long as costs are low. As costs increase, and especially when costs approach their limit, public concern for the environment will tend to diminish (and perhaps some environmental problems may appear resolved), and overall spending will tend to decline. Thus, the equation describing longitudinal change in spending for the environment can be expressed as

$$dZ/dt = XY(1 - Z) - Z. \tag{6.4}$$

In combination, equations 6.2, 6.3, and 6.4 constitute a nonlinear system of three interconnected differential equations. The three state variables, environmental damage, public concern for the environment, and spending for the environment interact longitudinally in a continuous fashion. As the system is currently specified, interactive cycling among all of the state variables is possible, as is demonstrated below. However, to capture a more complete characterization of the cycling between public concern for the environment and electoral activity (a generalized empirical reality that has been described more fully by Stimson [1991]), the above system requires a modification.

The modification is to allow the system to vary according to the partisan politics originating from the White House. This modification can be accomplished using parameter p in equation 6.2. The analysis below has two levels of sophistication. As a first approximation to modeling the interaction between politics and the environment, parameter p oscillates between two different values, depending on the ideological perspective of the current president. When a conservative president occupies the White House, that president's supporters are not likely to include environmentalists wanting large increases in governmental spending on the environment, and thus parameter p is likely to have a relatively low value. However, a more liberal president will be more closely tied to the desires of environmentally active constituents. In such a situation, the value of parameter p is likely to be relatively high. Thus, we will need to vary the value of parameter p in a systematic fashion in order to see how the change in presidential perspective affects overall environmental damage, given the other elements in the model.[4]

Oscillating the value of the parameter p between two discrete values is considered a first approximation to modeling the interaction between politics and the environment because many environmental policy changes that occur across administrations may happen more gradually. Discrete changes are examined first in order to identify causally the behavioral characteristics associated with specific components of the model in its simplest form. Later in this analysis, the model is extended to include gradual changes in the parameter p that would reflect less sudden policy alterations that may be more typical of many situations.

But let us be clear at the outset about what is being captured by the parameter p. There are two ways of thinking about the parameter. The first is that the government actually cleans up the environmental mess that is made by itself and others. However, a second motivation underlying parameter p reveals a more subtle reasoning.[5] Governments sometimes clean up toxic disasters. But, perhaps more commonly, governments actively reduce environmental degradation by phasing out dirty technology in favor of newer and cleaner technology. Examples of this can be found in the construction of power plants, the use of catalytic converters on cars, the removal of lead from paint, and in many other areas. Industry initially resists such conversions to cleaner technology, but once the conversion is made, the situation is relatively permanent. The reason for the permanence is cost. Industry does not want to invest in new infrastructure twice: once to clean things up and a second time (if allowed) to get dirty again.

The desire of industry to stay clean once it is forced to invest in cleaner technology does not nullify the reasoning behind the specification using parameter p. The reason is that society is always changing, and industry is always growing. Thus, there will constantly be new products and new industries that will be producing new sources of environmental degradation. The parameter p captures the government's overall ability to keep on top of this ever expanding and constantly changing problem. It is not just that government sometimes cleans up environmental damage. Government also inhibits, through a variety of mechanisms, the development of new environmental problems. Thus, a high value for the parameter p reflects a government that is actively engaged in developing and maintaining a cleaner environment through a variety of means. The phasing out of old dirty technology and the gradual introduction of cleaner technology is just one such means of the term "clean up." The parameter p captures the summary effect of the total efforts.

There is very substantial empirical evidence suggesting that varying the value of the parameter p (both discretely and gradually) is a useful approach to modeling the interaction between politics and the environment. Since public concern for the environment began to heighten in the 1970s, there have been marked contrasts across administrations with regard to environmental concerns. Wood (1988) as well as Wood and Waterman (1991) have demonstrated that agency leadership is the critical factor in determining the direction of administrative environmental activities. This affects dramatic and sudden shifts in agency funding as well as the long-term effectiveness of regulatory activity. Indeed, political appointments generally have greater influence over the government's effectiveness in dealing with the environment than any other single source of governmental activity, such as legislation, budget variations, and even actions by Congress. In sum, it is administratively orchestrated politics that determine whether or not an environmental agency will be an effective advocate of environmental concerns.

Two heuristic examples are useful here. In the early Reagan years, a determined—and generally successful—effort was made to quickly curtail the scope, enforcement, budget, and effectiveness of environmental regulation. The detailed story of these efforts, which included the very effective appointment of an antienvironmentalist attorney, Ann Buford (formerly Gorsuch), can be found in Wood 1988, Harris and Milkis 1989, Vig and Kraft 1984, Waterman 1989, and Wood and Waterman 1991. However, such visible efforts are not needed in order to subdue an agency. Late in the Bush administration, it was reported that the Justice Department systematically, but quietly, blocked the prosecution of environmental crimes, thereby fatally derailing the efforts of the Environmental Protection Agency to enforce its own regulatory activity (*New York Times*, 30 October 1992, A13[N]).

Thus, there is no one way to measure an administration's concern or lack of concern for the environment. If an administration wants to undermine environmental regulatory activity, there are many ways to do this. Moreover, the same is true if an administration seeks to be broadly supportive of environmental concerns. It is for this reason that parameter p is specified as a summary estimate of all current environmentally related activity that reverses total levels of degradation, assuming that each administration will use whatever means are currently at its disposal to achieve its desired goals.

The Numerical Experiments

As is typical of nearly all such nonlinear and interconnected systems of equations, it is not possible to uncouple the separate state variables in order to evaluate the system's behavior using techniques of indefinite integration. Moreover, linearizing techniques are both cumbersome and not very useful in the current situation. Rather, to learn from the system described above, it is best to approach the system numerically with Hamming's maxim constantly in mind, "The purpose of computing is insight, not numbers" (Hamming 1973, 3).

The strategy of the current approach to analyzing the above system is similar to the way one would approach any empirical problem. First, a baseline description of the system is established. From this baseline description, the general behavior of the system in its simplest form is identified. Subsequently, experiments are performed in which one change is made to the system at a time, thereby gaining an understanding of the effect of that one influence on the overall model.[6] In this way, the model in its baseline characterization acts as the controlling specification with regard to the component that is being varied.

Figure 6.1 presents a phase portrait representation of a sample trajectory for the model.[7] This figure is used here as the baseline characterization of the system's behavior. The three axes represent the three state variables in the system, environmental damage, public concern for the environment, and environmental cleanup costs. In figure 6.1, the parameter p is held constant

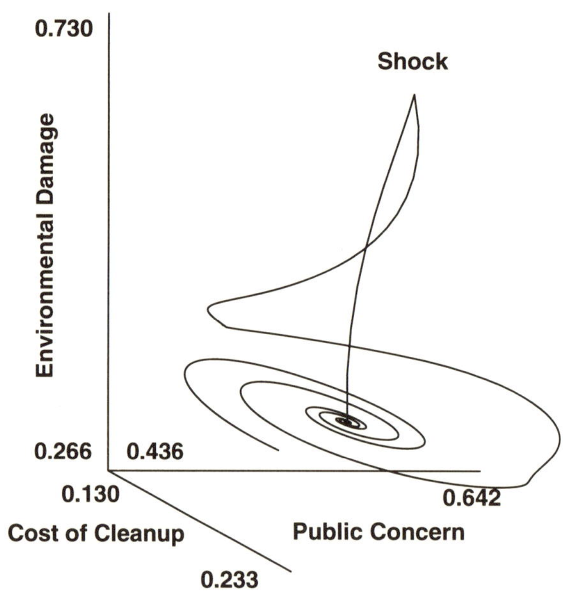

Fig. 6.1. Phase portrait representation of environmental damage, public concern, and environmental cleanup costs: sample trajectory with six-year public lag and *no* partisan policy differences. All trajectories spiral to a single global attractor, thus moderating environmental damage, despite short-term catastrophes.

with a value of one, and is therefore not varied between two values.[8] This represents a situation in which there are no policy differences between the parties, and all presidents behave similarly with regard to favorably promoting environmental policies. Moreover, the lag in public awareness with regard to concern for the environment is held at a moderate value of six years for this simulation.

Note that the trajectory spirals into a stable equilibrium point from its initial condition (labeled "starting condition"). Moreover, this equilibrium point reflects a relatively low level of overall environmental damage, as would be expected in a situation in which both parties favorably support environmental policies. After the trajectory's arrival in the proximity of the equilibrium point, a large shock is added to the system to see how quickly the trajectory will return to the neighborhood of its equilibrium. This shock represents what would happen to the system following a short-term ecological disaster. Note that the return trip is relatively rapid.

Thus, the system as represented in figure 6.1 (i.e., in the absence of partisan policy differences and with a moderate lag in the public's perception of environmental damage), supports (at equilibrium) a stable low level of environmental damage with moderate levels of public concern and relatively low financial costs. The swings encountered while arriving at that point may seem a bit rough, particularly following a major shock to the environment. But the eventual arrival at a pleasant and sustainable equilibrium does not take long. All this makes sense and would be expected from a properly functioning representation of this process under the given conditions.

Partisan differences in environmental policies are added to the system in figure 6.2. Moreover, that is the only difference added with respect to figure 6.1. Thus, in figure 6.2, the parameter p was allowed to oscillate between zero and one every eight years. This represents a situation in which each party would hold substantial differences in environmental policies and in which each party would rotate their control of the White House every eight years.[9] A value of zero for parameter p would reflect an administration that, say, strongly supports economic growth over environmental protection. Thus, governmental actions to decrease further damage to the environment would be minimal at best. A value of one for parameter p would reflect the reverse, that is, an administration that actively promoted proenvironment policies.

In the situation given in figure 6.2, note that a large stable orbit develops in the system. Moreover, even the ecological shock to the system disturbs the trajectory from this stable orbit only briefly. Thus, in the long run, the model is quite sensitive to environmental policy variations between the parties that rotate in their control of the White House.

From an environmental and cost point of view, the model's behavior as represented in figure 6.2 is not good news. The problem is the large magni-

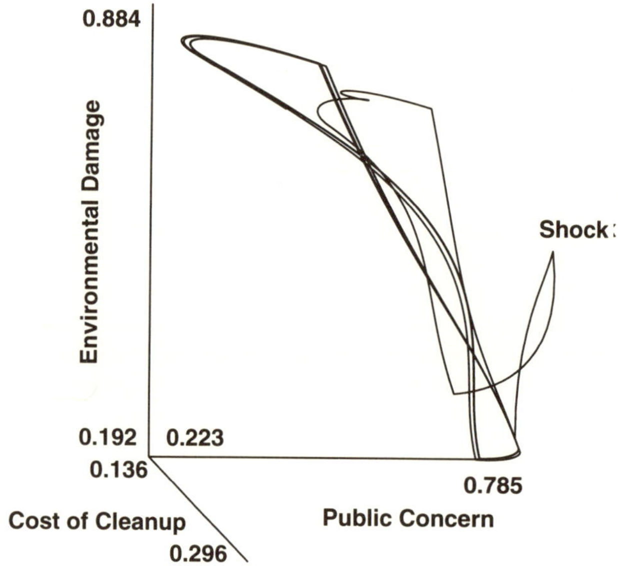

Fig. 6.2. System phase portrait: sample trajectory with six-year public lag *and* partisan policy differences. With a public lag of six years and political party differences, a stable orbit develops that leads to large, and potentially disastrous, changes in environmental damages as well as cleanup costs.

tude of the cycling and the consequent high level of governmental financial costs. When, say, a projobs president is elected, the previous environmental policies of his or her predecessor may have reduced the level of environmental damage. Moreover, with environmental damage more or less under control, the public's concern for the environment can begin to diminish. This leads to a relaxation of proenvironmental legislation and regulatory activity with its consequent eventual increase in environmental damage. At this point, the public desires change, a new party occupies the White House, and the process of reducing damage to the environment repeats.

One major problem identifiable in figure 6.2 is that the overall costs of this cycle appear to be much greater than the relatively moderate and stable costs represented in figure 6.1. From an environmental point of view, the repeated and high levels of damage are the problem. But from a financial point of view, it is the cycle itself, with its large and costly swings into the expensive programs needed for environmental recovery, that is the most damaging. Indeed, from both points of view, it might seem a better choice to reduce such long-term costs as well as environmental damage by avoiding the cycling.

But differing party policies with regard to the environment are not the only thing that can cause large magnitude and costly swings with this system. Indeed, even in the complete absence of political differences, the system is sensitive to the length of the lag in the public's perception of environmental damage. This sensitivity is demonstrated in figure 6.3.

In producing figure 6.3, the only thing changed from that used to produce figure 6.1 is the length of the lag in public concern regarding environmental damage. Policy differences between the political parties are not included in the computation of the trajectory in this figure. Here, the lag between environmental damage (e.g., the release of pollutants) and the public becoming concerned about the damage is 15 years. The result of this increased lag is the system developing a large stable orbit. This is a substantial characteristic change from the stable equilibrium that dominated the trajectory in figure 6.1. It is clear in figure 6.3 that the trajectory slowly converges to an orbit in which there are large changes in environmental damage, public concern, and economic costs. This is despite a political situation in which both political parties pursue similar environmental policies![10]

When political policy changes occur together with an increased lag in the

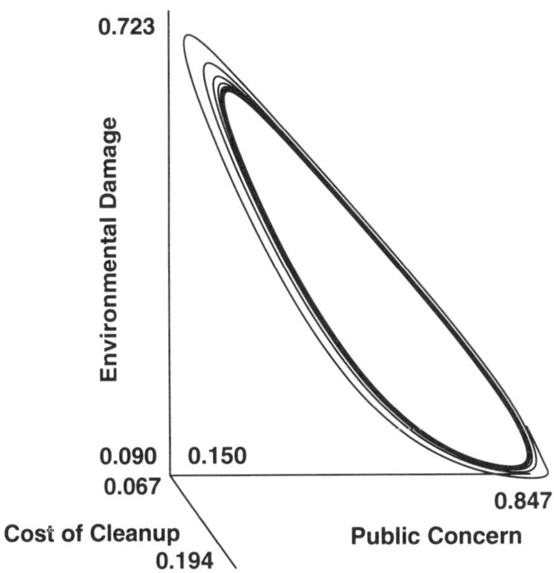

Fig. 6.3. System phase portrait: sample trajectory with 15-year public lag and *no* partisan policy differences. An increased lag in public concern produces large swings in environmental damage, despite no political party policy differences.

public's concern for environmental damage, the oscillations continue, but with more abrupt changes in direction and a much more complicated orbital form. Figure 6.4 combines both the longer 15-year lag as well as 8-year policy changes in the White House. From the phase plane perspective, the orbit in figure 6.4 seems to "flap" at lower levels of environmental damage, somewhat mimicking the swings of a pendulum. From this figure it is clear that directional changes in environmental damage can be much more radical than those associated with figure 6.3. Of course, these changes are associated with substantial changes in costs. But the costs often vary in this figure out of phase with changes in the other variables.

This temporal disconnectedness between costs and actual damage is directly tied to the interactive influence of the long lag in the public's awareness and the discontinuities inherent with the political policy differences. Moreover, this type of lagged cycling may be common. Notably, such cycling has been observed between the public mood and partisanship by Stimson (1991, 93–94) using a large body of survey data.

In terms of a goal of maintaining a healthy environment, this latter situation could be quite dangerous. The overall impact on the environment is

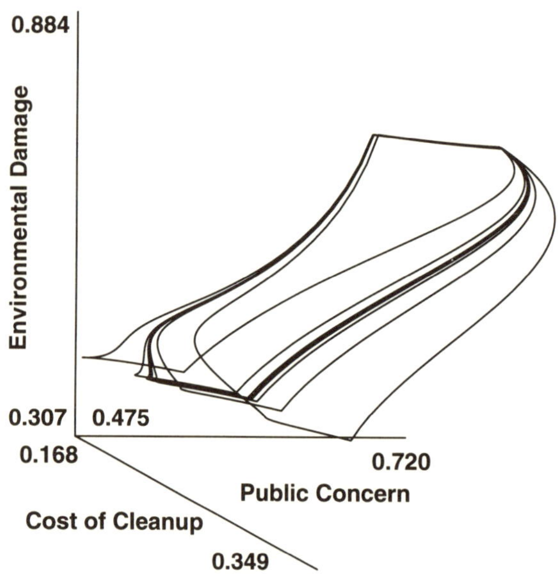

Fig. 6.4. System phase portrait: sample trajectory with 15-year public lag and partisan policy differences. The combination of an increased lag in public concern for environmental damage and political policy changes produces a "flapping" oscillatory orbit.

one of rapid departures from previous states and thus an increase in the longitudinal volatility of environmental quality. In some respects, this situation may partially reflect the general characteristics of the contemporary state of much environmental management in the United States today.

Gradual Policy Changes

In its present form, the model presented here assumes that governmental policy changes with regard to the environment are sudden. In some, and perhaps many, situations, this may not be an unreasonable assumption. Wood (1988) and Wood and Waterman (1991) have offered empirical evidence suggesting that national environmental policy changes are primarily influenced by the nature of top political appointments rather than by legislation, changing budgets, or Congress. Moreover, these policy leaders can display a remarkable ability to produce a wide range of change very quickly.[11]

However, not all policies—environmental or otherwise—will change suddenly. Indeed, more gradual changes across a broad range of policies may likely be closer to the norm of regulatory politics. At first, this may seem like a good thing. Gradual changes enable governments to come to grips with the consequences of previous policies and thereby to "fine tune" subsequent policy developments. Thus, we would expect increased predictability and a reduction in environmental instability when policy changes are more gradual. This, in turn, promises a reduction in the overall level of environmental damage. Surprisingly, these intuitions may be largely false. Indeed, gradual changes in policy need not eliminate dangerous and large-scale oscillations in environmental damage, and they can also end up hindering our overall ability to manage the environment.

To show why we are not necessarily better off with gradual policy changes with regard to the environment, it is necessary to reformulate the model to include such gradual, rather than sudden, changes. Thus, we need a new equation in the system. This equation will structure change in the parameter p. As described above, the parameter p is the government policy response to environmental damage. A high value for p suggests that the party in power is strongly engaged in reducing environmental damage through regulatory and other activities, whereas a low value for p indicates the opposite. Previously, the model has allowed p to flip from one value to another, thereby reflecting discrete changes in policy preferences across administrations.

We now want to vary p continuously rather than discretely. That is, the value for parameter p must change continuously in the direction of a desired partisan goal for that parameter. For example, a Democratic administration may desire p to have a high value whereas a Republican administration may think it more ideal if p were at a lower value. These desired values are partisan goals, to which the value of the parameter p should asymptotically approach depending on which party is in power at a point in time. Algebraically, this is accomplished by introducing into the system equation 6.5.

$$dp/dt = ep(g_{Dem} + g_{Rep} - p) \tag{6.5}$$

In equation 6.5, g_{Dem} and g_{Rep} are the partisan goals for the parameter p. Only one such goal can operate at a time since only one party can control the presidency at a time. Thus, when a Democratic administration is in power, g_{Dem} is set to its ideal value and g_{Rep} is set to zero. The reverse is true when a Republican administration is in power. Thus, all of the terms in parentheses in equation 6.5 act in combination as a logistic directional control for the evolving value of parameter p. For example, when a Democratic administration is in power, the value of p changes in the direction of g_{Dem}. Change slows down when p approaches g_{Dem} as a limit. The parameter e identifies the rate (proportional to the current value of p) at which this movement in the value of p takes place. All of this changes p from a constant parameter to a variable parameter. The entire model is now a four-equation system which is collected and summarized below.

$$dX/dt = rX(1 - X) - pXY - kX$$

$$dY/dt = X_{old}(1 - Y) - Z$$

$$dZ/dt = XY(1 - Z) - Z$$

$$dp/dt = ep(g_{Dem} + g_{Rep} - p)$$

Again, the variables and parameters are

- X: environmental damage
- Y: public concern for the environment
- Z: economic costs of environmental cleanup
- p: governmental policy response to environmental damage (a variable parameter)
- g_{Dem}: ideal Democratic policy response goal
- g_{Rep}: ideal Republican policy response goal
- r: pollution growth rate parameter
- k: natural pollution decay rate
- e: parameter determining speed of government policy changes toward partisan goals

Our first test to see what effect this change in determining the value of parameter p has on the overall behavior of the model is to duplicate all of the other conditions used to create figure 6.4, which included partisan policy differences as well as a 15-year lag in public concern for environmental damage. This new simulation is presented here in figure 6.5.

In figure 6.5, the values of parameters r and e are set equal to one, and the Democratic and Republican ideal policy response goals (g_{Dem} and g_{Rep}) are set equal to one and zero, respectively. Thus, the situation is very similar to that of figure 6.4, except that the value of parameter p is allowed to move continuously and gradually between two oscillating limits. From figure 6.5, the primary effect of this change on the overall model is to round off the sharp edges of the trajectory that were previously the result of sudden policy changes. Large magnitude oscillations in environmental damage still occur, and the flapping oscillation pattern of figure 6.4 reappears, albeit in the form of rounded loops rather than more sharply angular movements.

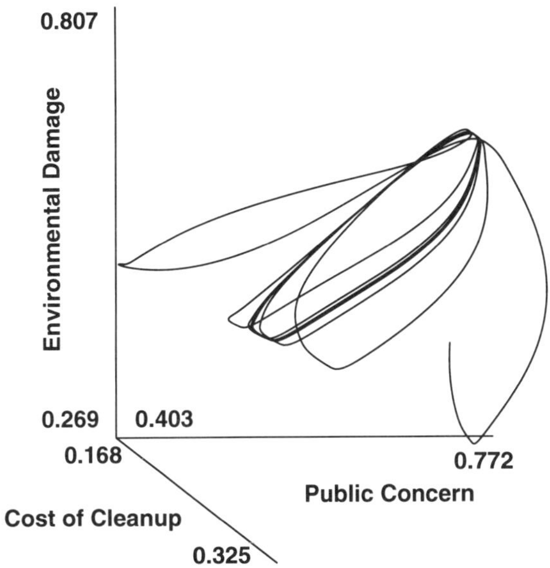

Fig. 6.5. System phase portrait: sample trajectory with 15-year pubic lag, partisan differences, and gradual partisan policy changes. The combination of an increased lag in public concern for environmental damage and gradual political policy changes can produce smoothed versions of the flapping oscillation observed in figure 6.4. Here, r = 1 and e = 1. Democratic and Republican ideal policy response goals are 1 and 0, respectively. This corresponds to position A in figure 6.6.

A natural observation relating to figure 6.5 is that we are looking at only one value for the parameter *e*. Recall that the value of parameter *e* controls the rate at which government policy changes toward the partisan goals emanating from the White House. High values for this parameter correspond to rapid movement toward these goals. Very high values can produce movement that mimics that associated with sudden shifts in policies, as portrayed in figure 6.4. Very low values make movement between the two goals sluggish at best.

At this point in the analysis, what is needed is a way to portray the overall behavior of the model through a continuous and realistically suitable range of values for the parameters *e* and *r* without paging through a countless series of trajectory phase portraits. This is important since we do not yet have one hundred or so years of data from which to estimate these parameters. One counterargument to the current analysis would be that the particular values used here for these two parameters are not likely to be the exact values that the system will actually have once it is estimated, given an appropriately long collection of time series data. Other parameter values may produce less dangerous systemic behavior. Moreover, it is likely that these parameter values will migrate over time (i.e., not remain constant) to keep pace with evolutionary changes in our social and political cultures. Thus, it is important to describe the general topology of the environmental playing field on which the values of these parameters are placed so that one can generalize about the overall characteristics of the system. As will become obvious below, one can find little solace in the idea that it may be better to wait to collect the data before worrying about the intricacies of the political-environmental connection. Indeed, in terms of environmental risk and a desire to engineer an effective collection of environmental policies, it may be difficult to imagine a less hospitable general setting.

To do this, a new measure is needed that captures two aspects of the overall behavior of the model. The first is the sum of the absolute value of total change in environmental degradation. Small values of this aspect indicate that change is slight, and thus probably more manageable from a regulatory perspective. On the other hand, large values of total change indicate dangerous levels of environmental volatility. The second aspect of our new measure must indicate how much directional change there is in the model's behavior. This is captured by counting the total number of changes in sign in the model's four derivatives for a fixed period of time.

The new measure is calculated by weighting the total magnitude of change in the environment by the total number of sign changes in the system's derivatives. This measure is an estimate of overall environmental turbulence. High levels of environmental turbulence indicate large magnitudes of overall environmental damage combined with frequent oscillations in the direction of change. Low levels of this measure indicate relative stability in environmental damage, both in terms of the total magnitude of change as well as the direction of movement that does occur. The measure is comparable across trajectories as long as the length of time for all trajectories is held constant.

Figure 6.6 is a portrait of the measure of environmental turbulence across a continuous range of values from zero to two for the parameters *e* and *r* (the pollution growth rate parameter).[12] The environmental turbulence measure is shown on the third dimension of the figure by the shading. Brighter shades in the portrait indicate high levels of environmental turbulence whereas darker shades indicate lower levels of turbulence. The letter "A" on the figure indicates the location of the parameters used to construct the trajectory presented in figure 6.5.

In figure 6.6, note that continuous change in the values of the parameters does not produce gradual changes in the measure of environmental turbu-

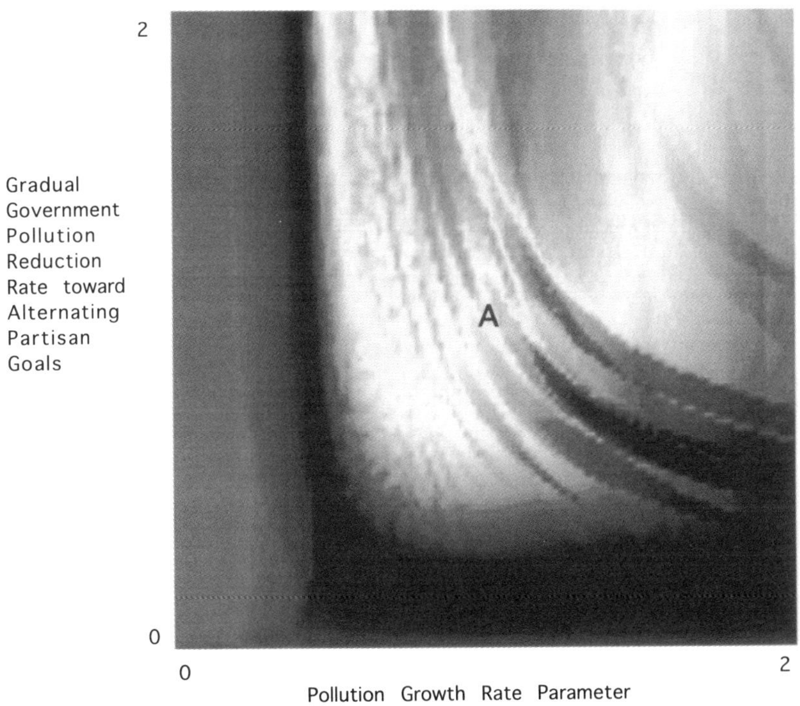

Fig. 6.6. A portrait of the total magnitude of change in the environment weighted by the number of sign changes in the system's derivatives. Brighter shades indicate large magnitude changes in the quality of the environment combined with many shifts in the direction of change for the variables, i.e., high levels of environmental turbulence.

lence. Indeed, the portrait in figure 6.6 is filled with closely interwoven variations in shading, indicating rapidly changing levels of turbulence. That is, turbulence does not increase smoothly as, say, industries increase their level of pollutant discharge. Nor is the change in turbulence gradual if the government pollution reduction rate changes evenly. Moreover, the only areas in figure 6.6 in which darker shades dominate in a significantly continuous fashion are for situations with very sluggish rates of partisan policy change (i.e., very low levels for the parameter *e*) or under conditions of very low levels of environmental damage, neither of which seem to realistically describe current regulatory or environmental conditions.

One of the important lessons to gain from figure 6.6 is that controlling the environment under arguably realistic conditions with regard to our current approach to environmental management may not at all be easy. In order to manage the environment, one must be able to predict what will happen when changes are made. The results of figure 6.6 suggest that this may not be easy, or even possible given current measurement inadequacies. Attempts to manage the environment within a setting of significant differences in partisan approaches can lead to large-scale and highly volatile environmental changes, many of which may not be easily reversible from an ecological point of view.

As bad as this sounds, the situation is worse. There are large areas in the parameter spaces of parameters *e* and *r* for which the determination of what will happen next in the environmental trajectory is nearly impossible to predict. This is due to complexities in the oscillatory components imbedded in the trajectory movements. The oscillations portrayed in figure 6.5 seem easy to follow; but for other values of the parameters *e* and *r*, the situation changes dramatically.

To show these conditions, one additional measure is needed. This measure must characterize the oscillatory complexity of the trajectories in two ways. It is important to know the magnitude of the oscillatory components as well as the variety in these same components. To do this, it is necessary to calculate a Fourier series, and its associated periodogram, for the environmental measure. From this series we obtain two quantities. The first is the total power of the series, indicating the overall magnitude of the combined oscillatory components.[13] The second is the standard deviation of the periodogram elements for the nonzero positive frequencies. This quantity characterizes the amount of variety in the oscillatory components as determined by their frequency. Thus, the measure of the longitudinal complexity of an environmental trajectory is computed by weighting its total Fourier power by the standard deviation of its periodogram elements. High levels of this measure indicate greater levels of oscillatory complexity in the trajectory whereas lower levels indicate rather predictable oscillations.

Figure 6.7 presents a portrayal of the model's longitudinal complexity

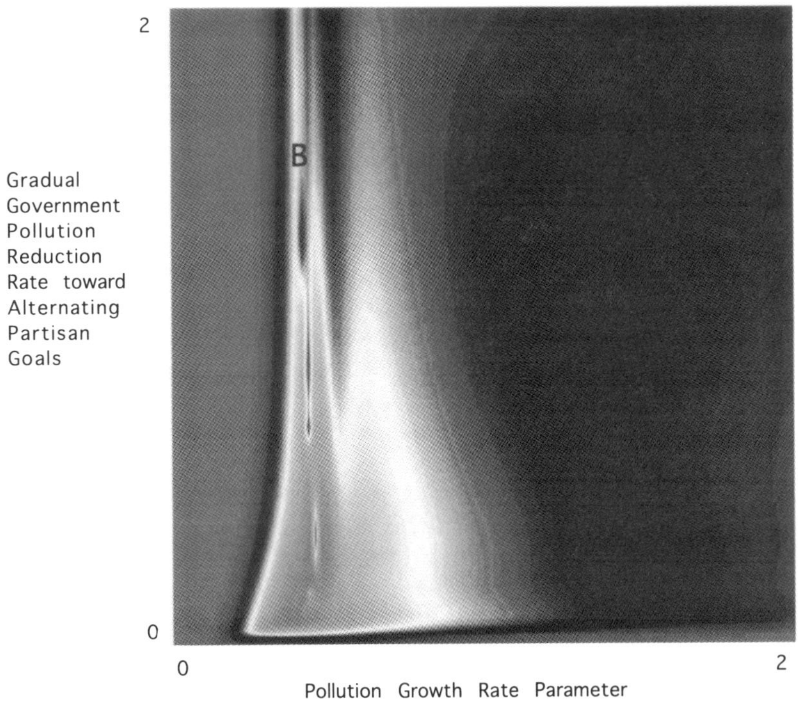

Fig. 6.7. A portrait of the total Fourier power of the four-equation environmental system weighted by the standard deviation of the periodogram elements for nonzero frequencies. Brighter shades indicate greater levels of longitudinal complexity in environmental change, indicating greater difficulty in predicting—and thus controlling—environmental damage.

over the same range of parameter values as used in figure 6.6. Brighter shades indicate higher levels of longitudinal complexity whereas darker shades indicate the reverse. The "flamelike" bright areas to the left of the figure mark an area of the parameter space in which the longitudinal complexity of the oscillatory components of the model's trajectories is very complicated. The letter "B" in the figure marks the location of the parameter values used to construct the example trajectory presented in figure 6.8.

From figure 6.7 it is clear that there is an approximately vertical "passage area" in the parameter space of the model in which the model's longitudinal complexity increases dramatically. Moreover, this passage area of greater complexity is not of uniform width as one travels from left to right (along the axis of the pollution growth rate parameter). Moreover, variation in the level of longitudinal complexity is often quite large for very small changes in parameter values. (For example, observe the closeness of dark and bright areas near the letter "B").

Trajectories with higher levels of longitudinal complexity appear markedly different from that presented in figure 6.5. Figure 6.8 presents such a complex trajectory using the parameter values e = 1.7 and r = 0.5. (Again, this corresponds to position "B" in figure 6.7.) From figure 6.8 it is clear that policy managers would have a difficult time predicting the future of environmental change given this political and pollution context. The oscillations are evident, but they do not settle down into a clearly discernable pattern within any reasonable length of time. On a technical level, such complexity in trajectory structure may not be certifiably chaotic, classically defined, since the characteristic Lyapunov exponents do not indicate a strong sensitivity to initial conditions (an analysis not shown here). Nonetheless, a Fourier analysis using the current parameter values reveals a complex periodicity very comparable to that of chaotic systems. Yet from a macro substantive perspec-

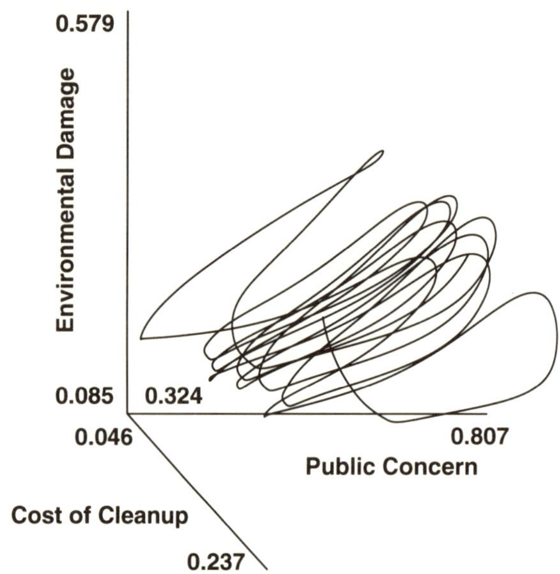

Fig. 6.8. System phase portrait: sample trajectory with 15-year public lag, partisan policy differences, and gradual partisan policy changes using alternate parameter values. Modest parametric variation in the four-equation model can produce very complex oscillations in environmental damage. Here, r = 0.5 and e = 1.7. Democratic and Republican ideal policy response goals are 1 and 0, respectively. This corresponds to position B in figure 6.7.

tive, there seems little difference between these complex oscillations and pseudorandomness since the level of longitudinal complexity is sufficiently high to make a practical ability to predict the future virtually impossible.[14]

Substantively, this implies that a great deal can happen quickly with regard to the environment when small changes occur in pollution and policy rates and when the political parties differ in their environmental policies. This is precisely the fear raised by Schelling (1992, 8) with respect to global warming. It is this potential characteristic of the environmental system that holds the greatest long-term danger for planetary ecological management.

Conclusions

It would be absurd for an astronaut in a space vehicle to start a fire in the cabin to keep warm. No one would question this because we understand that the regulatory mechanisms within the vehicle would be overloaded from the consequences of the fire. The problem with making a similar statement with regard to damaging the environment of our planet is that we do not entirely understand the automatic processes of environmental management. These processes include political structural components as well as ecological and physical components, and these political structural components have been largely ignored in the environmentally related literature.

It has been the purpose of this essay to describe some of the political components that may strongly interact with the rest of the overall biological system. The normative hope is that the discussion will engender a more balanced approach to the discussions of our environmental problems. The social scientific angle to the environment is not just economic. Since politics is at the root of all attempts at environmental management, political scientists should play a major participatory role in the evaluation of current problems and in the prescription of current and future remedies.

This analysis presents a model of environmental change in which the political, social, and economic inputs into the system interact nonlinearly to produce highly varied global patterns of ecological damage. The model is developed with respect to presidential electoral and regulatory practices in the United States. The results of the analysis suggest that relatively minor parametric changes in the system can lead to major alterations in the longitudinal patterns of environmental change.

One of the major implications of these experiments is to suggest that the regulation of our environment may be a much more challenging task than we have presently envisioned. Indeed, our current electoral structures may directly assist the long-term degradation of our environment. The complexities that exist within the simplified system investigated here suggest a greater level of complexity in the actual physical system. The political components of that complexity certainly play an important role in the determination of the general structure.

In terms of recommendations, it is likely that substantial benefit for environmental management from a political point of view would be to reduce the impact of party policy differences on ecological cycling. A complete plan on how to do this is beyond the scope of the current analysis, since the purpose of this discussion is more to demonstrate the dilemma we are in than it is to offer a way to get out of that dilemma. However, some initial suggestions to help initiate the discussion of solutions may be helpful.

Firstly, it is useful to observe that the Federal Reserve Board in the United States was developed in order to deal with potential cycling problems in the economic arena. Thus, we have experience with related problems in other areas and the effectiveness of some of the solutions that have been attempted. It is not likely that a simple political compromise between the political parties would last sufficiently long to be meaningful to the long-term global properties of the system. The temptations to exploit short-term political gain would eventually destroy any temporary effects that would result from the compromise. Something more permanent and substantial is needed, and the creation of some type of independent "Environmental Preserve Board" may be one answer.

It may also be possible, perhaps more so in countries other than the United States, to create an elected position through constitutional means that is independent of the influence of the office of president or prime minister. This elected official would serve as the head of the environmental regulatory agency of a given country. This could have major implications to the management of the environment, since the official would gain reelection solely on the basis of whether voters were happy with how the environment is being regulated. Thus, it would not be possible to avoid the issue of the degradation of the environment by talking about other issues in which the voters may also be interested. The basic problem is to isolate the environmental concerns from other political matters so that the one does not get lost in the soup of the other.

Yet another suggestion would be to institutionalize the role of the United Nations in monitoring, and to some extent regulating, the planetary environment. Rules could be constructed that would establish planetary norms with which each country would have to abide. Some fair system of penalties would have to be established in order to obtain cooperation across nations. Such penalties would likely involve trade and economic issues, but technology transfers and political cooperation in other areas could also be factors. Interestingly, the United Nations already seems to be becoming much more involved in other global matters since the end of the cold war. Thus, the stage may already be set for the emergence of a new environmental role for this organization.

But it is important to understand that helpful environmental management cannot function only on the level of bureaucratic organization. At base, individual humans must understand the need for this management. In particular, the regulation of the environment is influenced by the lag in the public's perceptions of our environmental problems. To some extent, this can be addressed through an enhanced role for environmental education in our society. On the level of formal education, it is most likely that a comprehensive approach to curriculum reform across all age categories will be required in order achieve any lasting benefit to the overall problem, since generational biases often persist for lifetimes and it may be too late in the game for college courses to make a meaningful impact.

This last recommendation cuts to the heart of the matter. If nonlinearities dominate the political-environmental system, it may be futile to try to fine-tune current environmental policies in a rational decision-making sort of way in an attempt to fix the ecological problems. If political leadership, regardless of party, tends (at least on average) to reflect in practice the policy preferences of large numbers of a nation's citizens, then it is necessary to upgrade the preferences of the citizens through education before environmental disaster does it for us. It is not that evil political and corporate leaders continually dupe the ignorant political masses. Rather, in a democracy, it is the masses that let their leaders do what they do. At base, unless large majorities of the citizens of democratically governed countries become strongly environmentally sensitive, there can be no long-term solution to this problem. The electoral and ecological cycling will continue as parties with differing and large constituencies oscillate in office, and the costs of environmental cleanup will remain both high and volatile. Possible governmental/structural fixes may help in the long run, but only if the educational component is also present.

Fundamentally, any successful approach to environmental management will require an understanding of the political processes that influence that management. At the outset, it is important to understand how these processes extend to the basic electoral practices of our democracy. The current discussion has attempted to demonstrate that the political-environmental connection within the context of contemporary democratic governance is a complex relationship at best. This nonlinear complexity is associated with questions relating to human quality of life, and indeed, survivability. I address these matters not as mathematical curiosities, but as issues with a direct bearing on the physical characteristics of our planetary habitat.

CHAPTER 7

Toward a General Theory of Nonlinear Political Evolution

Complex Systems

This volume has focused on nonlinear political phenomena across a variety of settings. There is no one place to look for nonlinearity in human existence. Indeed, the more general rule should probably be to expect nonlinearity everywhere and to make note of those few exceptions in which human behavior may actually be linear. Nonlinearity itself, however, is a characteristic of complex systems, and I have here investigated such systems in four important substantive areas: an electoral landslide in the United States, the collapse of a fragmented electorate in the Weimar Republic, the interdependence between our system of governance and the destruction of the planetary ecosystem, and the relationship between individual decision making and social context. All of these complex systems have general properties, and it is important to develop a classification scheme for such properties so as to better place the current contribution within a broader framework.

Taking initial guidance on a classification of complex systems from Crosby (1987), all systems—including social systems—are evolutionary. This is a consequence of the fact that if one goes back far enough in time (to the Big Bang, if necessary), no current system existed; thus all have evolved. At the most abstract level, there are three types of systems: physical laws, genetic transfers, and decision making. Physical laws concern processes that involve physical and chemical systems, including the formation of such things as galaxies, mountains, winter storms, and rain. Genetic transfers govern the biology of living systems. Decision-making systems are based on thought, and are the primary factor in the evolution of human social systems.

Within each of these systems, there are evolutionary processes. Four separate processes may occur within any one system, although generally not simultaneously. These are regular, periodic, chaotic, and catastrophe. It is important to note that I have identified all of these nonlinear processes within realistic political settings in this book.

Regular processes follow an evolutionary path that begins with a period of growth, followed by the intersection of a bifurcation, and finally a constant steady state. The growth process is typically a positive feedback situation in which a system's output induces a further expansion of subsequent output. Since accelerated growth cannot continue forever in any system, this condition must metamorphose into some other state. The metamorphosis occurs when the system variables transit a bifurcation point or set of points.

The analyses in chapter four regarding the landslide election between Johnson and Goldwater are an example of dynamic processes that are predominantly regular. In this case, the steady state target of the trajectories are equilibria that are located within equilibria marshes, identified graphically in figures 4.2. and 4.5.

Bifurcation sets can affect all evolutionary processes, and they have played important roles in many of the substantive analyses presented here. Near a bifurcation set, system behavior can be highly unstable given relatively small fluctuations in the values of the parameters and variables. Instabilities can occur due to the system's own internal dynamics, or because of interdependencies with other systems that are evolving in their own way. With regular processes, the bifurcation simply marks the net shift in dominance between positive feedback and negative feedback elements, where the dominance of negative feedback leads to the rise of a constant steady state. This is an ideal characterization, however, and the reality is that there is no such thing as a permanently constant steady state with regard to social phenomena. Small-scale internal changes or externally mediated interdependencies eventually lead to change in the state, or perhaps relatively slow evolutionary movement (in contrast with the growth phase).

The analyses in chapter six regarding politics and the environment identified periodicity as a critical component influencing human management of the planet's ecosystem. In general, periodic processes collapse onto limit cycles, and figure 6.2 is a graphical representation of such a condition. It is my view that many other situations exist in political and social settings in which limit cycles play a crucial role in the relevant dynamics, and social scientists need not consider models such as the classic Lotka-Volterra predator-prey system as distant from empirical settings involving humans. Stimson's identification of periodicity in cycles involving the public mood in the United States is an excellent example in this regard (Stimson 1991).

Periodicity is not restricted to limit cycles, however, and these analyses have also encountered situations in which more complex periodic structures have dominated the dynamics of human behavior. Chaotic processes yield systems that mimic randomness in the period and amplitude of the state variables. The Lorenz equations are a suitable representative of this class of dynamics. Bifurcations play a central role in chaotic processes since they are linked to the phenomenon of period doubling in correspondence to parametric variation.

While the identification of what is or is not chaotic is not yet fully resolved in the literature on dynamics, some indicators can be used to argue persuasively for the presence of chaotic processes in many settings. In the current investigations in chapter six, dynamic processes are identified that are extremely complex, having no discernable resolution of a finite and noncontinuous spectrum of frequencies. These processes may or may not be certifiably chaotic, but they are complex in the sense that chaotic processes are complex, and this is what matters if we are to understand the problems we face sufficiently to have any hope of successfully managing the political environment.

In this book, catastrophe dynamical processes are identified in two noteworthy and very different political settings. The first involves the electoral success of the Nazis during the period of the Weimar Republic. The second is in the interaction between individual rationality and a voter's political milieu.

From a technical perspective, a catastrophe is associated with the sudden disappearance of an attractor together with its basin. Typically, a catastrophe links two relatively regular evolutionary processes. The processes are called "relatively regular" since there is no guarantee of a steady state situation for an appreciable period of time. Indeed, it is possible for the postcatastrophe dynamics to include a growth phase leading to another catastrophe, followed by more growth, and so on. Bifurcation sets are essential players in catastrophe processes, since they mark the boundaries of an attractor's influence.

All of the substantive examples of nonlinear dynamical processes investigated here also experience situations of system maintenance, either for a specific period of time, or occasionally with respect to a particular social context. System maintenance occurs when negative feedback dominates the dynamic processes such that the system is stable. In regular processes, this is marked by the arrival at the constant steady state. In other situations, maintenance can occur with regard to periodic and even chaotic processes. While a catastrophe process itself cannot be stable from a maintenance perspective, a repeating cycle involving intermittent catastrophe dynamics can be maintained. Also, a catastrophe process can lead to a regular process in which system maintenance is a crucial aspect of the subsequent dynamics.

Every situation examined in this volume is an example of a decision-making system. Moreover, decision-making systems involve all of the nonlinear dynamical processes described above. Decision-making systems do not have to be based on an assumption of rationality, since any cognitive structure, including stimulus response, will suffice. The essential aspect is that thought processes of some type be the critical link to change in the system variables. For example, nonlinearities in our political management of the planetary ecosystem may lead to a catastrophic approach to a dissipative structure in the metasystem of life. We need not even be aware of these nonlinearities for them to occur.

Social systems can be extremely complex in general. I must also point out that these complexities are likely to require social scientists to develop their own set of tools and concepts in order to evaluate realistic social settings. For example, the current investigations have encountered dynamical properties of human behavior that required the application of a variety of new ideas, such as definitions of an equilibrium marsh, of when a society votes in equilibrium, a catastrophe superstructure, and near chaotic processes. All of the processes outlined in this chapter—and given realistic portrayals in previous chapters—are relevant to virtually all aspects of social reality, and all are essentially nonlinear. That social scientists rarely talk in such process terms does not mean that these dynamics are alien to the human condition. But if we are to make a final break from the intellectual dominance of the existing linear paradigm, we need still to ask why nonlinearity *must* occur with regularity. Why can human behavior *not* be linear?

The Cause of Nonlinearity

Identifying nonlinearity as a norm does not explain why it exists in the first place. Indeed, identifying the underlying cause of nonlinearity is the single most important aspect to the future development of a full and general theory of nonlinear political evolution.

In its most basic manifestation, nonlinear dynamics occur when things interact. This does not imply that things must interact with other things, as when people talk together. Indeed, something can interact with its own history as well. But the key concept relating to human life is that people do not live in isolation from the world that surrounds them. The surrounding world is a product of the past behavior of each individual, and thus the collective evolution of society is due to the history of the current state of affairs as much as it is due to the activities of the present.

Consider the basic concept of exponential growth, the idea used by Malthus to explain and predict the future condition of the planet's human population. Algebraically, this is expressed as

$$dN/dt = aN. \qquad (7.1)$$

Here, growth is a function of the current level of population. Of course, change in the population is a function of the current size of the population. Using the more generalized concept of interaction as it is being developed here, the current population is interacting with its past, and even the functional linearity of equation 7.1 cannot eliminate this longitudinally nonlinear relationship with history.

A situation in which there would be no interaction with the past would be if the right-hand side of equation 7.1 was a solitary constant. This would be a truly linear condition both functionally and longitudinally. Certainly few social phenomena change in this fashion.

Functional nonlinearity adds a new layer of complexity to the description of the human condition. From the perspective of social theory, functional nonlinearity is inherent in the social embeddedness of man. Especially with regard to social systems, functional nonlinearity addresses the concept of group and individual identities. That is, are humans individually distinct from those with whom they live, or are they fundamentally connected, as in the case of cells being connected to the larger organism?

Since the answer to this question determines how we conceptualize society (and thus how we model it), it is crucial that we understand our own biases that may influence our scientific judgment. Oddly, the answer to the question may be culturally dependent. Different cultures view individuals differently. In the United States, individualism is a valued quality of human character. People like to express themselves as distinct from others by the way they look, talk, dance, as well as the with the kind of house they live in or the kind of car they drive. In other cultures, however, the reverse is true in many situations. For example, there is some degree of social truth in the Japanese maxim, "The nail that stands out gets pounded down."

Recently, there has been an attempt to rank cultures on the degree to which individualism or collectivism dominates social life (see the *New York Times*, 25 December 1990, B13 and 15[N]). The United Stated is typically found to be among the most individualistically oriented cultures. Countries like Peru, Thailand, Taiwan, and Venezuela are among the most collectivist cultures. Functional nonlinearity is a natural approach to modeling social change given a collectivist worldview of the human condition. This is a consequence of an understanding of human life as fundamentally interdependent with, minimally, other human life. Functional linearity, on the other hand, makes more conceptual sense given an individualistic view of people. In this latter case, characteristics of the individuals shape the way they behave. It matters less on which side of the tracks one lives; the ideas that are in a person's head—like whether they like certain social policies—are the things that really count.

I argue here that the social-theoretical hegemony of linearity is not only a matter of mathematical convenience. It is also a consequence of a cultural bias that emphasizes an individualistic view of human life over a collectivistic one. If individualistically defined characteristics are the determinants of social change, then correlational associations with various individual-level variables will satisfy the theoretical needs. Linear models can do this; it is their great strength. Moreover, it is only natural that individualism—and thus linearism—has dominated so much social empirical theory over the years given its origin in Western society.

Feedback, social interdependence, the attachment to history . . . all these things are characteristics of the various manifestations of nonlinearity. They cause nonlinearity to manifest, and they are inseparable from virtually all aspects of human social life. Descriptively, these concepts are used to help specify the complexity of human societies, and this is the real issue. Nonlinearity is a consequence of complexity, and it is our theoretical retreat from the identification of man's social complexity that binds us.

On Human Bondage

One of the great fears of Hannah Arendt is that we, as individuals, are becoming increasingly social. That is, human life is increasingly being determined by the requirements of the larger social body. Behavioralism in the social sciences is to be feared not because it incorrectly defines man from the perspective of a correlational statistic; statistics truthfully identify the behavioral characteristics of biological beings who participate in modern society. Her fear is based on the view that the behavioral sciences are too accurate in their description of the boundaries of human freedom individualistically understood (see Canovan 1974, 85–92). From her perspective, we are "a society of men who, without a common world which would at once relate and separate them, either live in desperate lonely separation or are pressed together into a mass. For a mass-society is nothing more than that kind of organized living which automatically establishes itself among human beings who are still related to one another but have lost the world once common to all of them" (Arendt 1961,89–90).

From our Western perspective, humans no longer are members of a clan, class, or family. Rather, they are components of a mass society, united by their nature, the commonality of which can be identified through the correlational breakdown of their individual characteristics. Behavioral social science rarely identifies the complexity of human existence as much as it categorizes individuals with regard to behavioral types. As strangely as it may seem at first, the dominance of the linear paradigm is a characteristic of this categorization process. The identification of slopes is the operationalization of this process.

Why then do we, as a society, press forward with our linear descriptions of modern life? Or, put differently, why do we retreat from the complexity of nonlinearity? The answer to these questions may be as profound as it is nonobvious. Could it be that we view the world through linear eyes because we fear the awareness of our lack of control over the nonlinear complexity of our own lives?

Look at the way we govern ourselves. Since the genesis of the governing system in the United States, for example, we have been frightened by the unpredictability of the future. If we are to believe Madison, the writers of the Constitution of the United States assembled a formidable defense against nonlinear evolutionary vagaries of the masses by setting up a federal system with checks and balances (in particular, *Federalist No. 10*). Although they did not use our modern terms, these foresighted theorists knew intuitively that the development of any society was inherently nonlinear, and that the complexity of that state of affairs had to be controlled in order to avoid evolutionary tendencies that were not considered favorable. Our governmental system is our attempt to hold a steady course amid a turbulent sea. In the language of dynamics, we seek to place a solitary constant on the right-hand side of equation 7.1. We seek a linear development that avoids falling into a basin of an unhappy attractor. We know these things can happen, and we fear them.

Examples of such things abound. Our Federal Reserve Board was created to avoid such unexpected nonlinear deviations in our money supply. We regulate our environment in an attempt to avoid singularities in potentially nonlinear dynamics that could end human dominance on this planet. We buy off our poor with minimal maintenance to avoid them burning our cities. Indeed, it is often claimed that Franklin D. Roosevelt so quickly established new social programs as an attempt to prevent a revolution in the country that could go either fascist or communist. We control our societies. We manage our individuals. We identify ourselves not by the meaning in our own lives but by the commonality in characteristics that we share with other subgroups in our mass society.

The acceptance of a new paradigm of nonlinearity destroys the mythical reasoning underlying how we control ourselves. It does so by not allowing us to bury the inherent complexity of human life with linear behavioral models that destroy the understanding of our reason for being. If we no longer shape our view of ourselves with the linear hammer of slopes, we must identify our inherent interdependence as individual contributors to a larger whole. Thus, the identification of our essential nonlinear nature does more than give us a richer bag of mathematical tricks with which we describe our societies. It encourages us to comprehend the true complexity of the evolution of our cultures, thereby freeing ourselves from the bondage of our mass society intellectualism that was Arendt's great fear that I mention above.

It is not that we are no longer beings with a society. There are too many of us to ignore the fact that we live not alone. It is just that we become free when we can see that which controls us. As long as we stick to a linear view of ourselves, we will continue to depart intellectually from a knowledge of that which we are, perhaps to avoid the awareness of our impotent control over our own destiny. We may try to control the dynamic vagaries of a nonlinear social existence by structuring the ways we govern our masses in the direction of linearized and metered change, and for long periods of time we may succeed. But we run from a faithful evaluation of our condition if we allow a rigid categorization process to structure our thinking as well. Our models become the language games of Wittgenstein, universes unto themselves, understandable only from within. No one so bound is free.

From this perspective, linearism is an intellectual attempt at control, though few see it for what it is. That it is blatantly untrue as a conceptual framework is nearly obvious by now. We adhere to its rigidity only at great cost to our own understanding of our essential nature. That we acknowledge our own nonlinearity just as we recognize nonlinearity in the rest of the manifest universe only concedes to our vital character. We live complex lives. When we understand how and why we live the way we do, we gain true freedom to determine our fate. Individualism and linearism are mutually reinforcing concepts in the social sciences. And from this point of view, an individualist ideology in the social sciences is primarily an imagined escape from the apparent imprisonment of complexity. But where is self-determination among the blind?

Society as a Living System

Interdependence is what makes our understanding of the complexity of society so challenging. Despite centuries in the development of modern mathematics, Robert May just recently informed the world that a relatively simple one-dimensional nonlinear difference equation could exhibit highly complex chaotic dynamics. Interdependence expands the level of complexity by increasing the dimensionality of our substantive problems. If our understanding of complexity in one dimension has recently grown, our understanding of interdependent social complexity is still in its infancy.

Interdependence addresses the idea that individuals are embedded within larger social units. Beginning with groups, the smallest of these larger units, interdependence occurs when individuals interact with other individuals. The group develops an organized structure, even if only minimal. For example, people listen while others speak, and certain individuals more heavily flavor the tone of the discussions than others. In a very real sense, a group is system, a living system. This is an important argument made by James Grier Miller (1978) with regard to the entire spectrum of human organizations. It is living because its members are a set of living organisms. The members of the group process both information, matter, and energy. They relate to one another face-to-face, and can do so for one extended period of time or during a number of interrupted sessions. Miller's argument is that groups, as well as higher level social organizations, are real living entities, although he, as well as I, acknowledges the debate on the subject (Miller 1978, 515).

From a biological perspective, the idea that groups, organizations, and society are living things is defendable, although with controversy. Biologists consider an ant nest to be one living organism. Individual ants are like individual parts in the body of, say, humans, in the sense that they differentiate their activities to perform the same functions as our body parts. The primary difference is that all ants are not housed under one skin. In my view, characterizing human organizations as living entities is a useful definitional approach to the identification of inherent interdependent complexity in our societies. But other scientists may not be so inclined to make this intellectual leap, and it is important to point out that the logic of the remainder of my argument is not lost as a consequence. We need not buy into the concept that groups are living in order to appreciate the added complexity that is added to the lives of people when one must consider that individual action is a consequence not only of the characteristics of individuals but also the result of everything that takes place at higher organizational levels. As individuals, we depend on these higher-level organizations. For example, I depend on my university for my salary, computer support, secretarial help, and countless other things. I also participate in university life. The relationship is reciprocal. Whether one considers a university a live entity does not change the fact that I would not be all that I am today outside of the university environment. It evolves with me, and I with it. We are interdependent.

It is one thing to describe my own personal growth, and another thing entirely to model my evolution jointly with that of my university. It is a still greater challenge to model the joint evolution of myself, the university, the professional organizations that I belong to, my friendship groups, the city within which I live, my state, my country, and my planet. The interdependencies expand and overlap, and it is not possible to model all interdependencies simultaneously. As a theorist, one has to choose the level of complexity that can be modeled with some degree of parsimony in order to investigate the more fundamental aspects of a substantive problem. Viewing human activity in terms of a nonlinear social system simply expands the level of complexity that can be captured in a particular model by formalizing at least some of the relevant interdependencies.

Minimally, this volume presents a wide variety of substantive areas in which modeling the nonlinear interdependencies inherent with our lives can yield useful results. My argument throughout has been that linear models represent too simple a representation of that complexity. I have never argued that researchers should completely abandon linear models. My assertion is that linearism has dominated our thinking process to an extent that more realistic representations of the human condition have not been realized. It is not that we need to stop using linear models. Rather we need to expand our understanding of the complexity that conditions our existence to include nonlinear perspectives as well. This means that we need to recognize that nonlinearity is an essential characteristic of our nature as living beings. We are nonlinear because we live complex interdependent lives. Moreover, our lives are built upon the historical lives of others. Thus, our inherent complexity extends through time as well. The goal of modeling the interconnectedness of all of this is ultimately an understanding of the cultural evolution of our species. No one nonlinear model can accomplish all of this by itself. But without a richer collection of attempts at understanding our integral nature, can we ever hope to know who we are?

We wish not only to understand ourselves, but to determine who we will become. Politics is our society's nervous system. It is how we communicate our disparate needs to the collective whole in order to shift resources and resolve our troubles. But we engage in politics not simply to solve society's accounting problems. There is a higher goal to our collective activity. It is my sense that we wish to evolve collectively into that which is better than our current state, even though many people may not be entirely conscious of this. In terms of specifics, we wish to eliminate hunger, disease, crime. But the elimination of negatives is not the goal. Nor is the goal the attainment of individualized prosperity. This book is written from an intuitive sense that at some inner level, most—and maybe all—humans desire to live as members of a society that is productively evolving. For example, we see this when we claim pride at our country's advances, in whatever field they may be achieved. And perhaps the day will come when we look down from the skies and reflect with pride on how we, as humans, have managed our homeworld.

Ultimately, we need to understand the nonlinearity, complexity, and interdependencies of our lives so that we are no longer passive riders on an evolutionary voyage that touches every level of our existence. Breaking away from an intellectualism that limits our ability to perceive our interplay with the surrounding universe is a necessary first step to achieving self-determination as a species. Intellectual linearism too narrowly restricts our view. To be in control of our own fate, we need to see clearly who we are and how we came to be. Of this I am certain: our current state of affairs is as nonlinear as the evolutionary path that brought us to where we are today. I can conceive of no intermediating factors that could alter the unitary probability that nonlinearity will structure our individual and collective human destiny as well.

Appendix

Some readers may be interested in the precise methods by which the models in this volume were estimated. For this reason, I have prepared a complete version of the estimation program that I actually used in conducting this research. The program can currently handle up to 30 parameters for nonlinear models with data structures similar to those encountered in this book, that is, two time points with many cases. The program is thoroughly annotated, and it should be possible for anyone with rudimentary programming skills to master this code in a reasonable amount of time. (I give my graduate students one month to work on it in their spare time during the summer.) The program is written in SAS IML (the matrix language of SAS), a package that is available in nearly all university settings in North America and Europe. Readers who want additional written guidance on the algorithms used here can consult my text on the subject (Brown 1995; see also app. in Brown 1991).

The essentials of the program can be transported to virtually any other language as well, although I suggest sticking with a matrix language that does not require subroutines to be external to the main program file. (Versions of Matlab through 4.0 have this problem. Version 5.0 may resolve this.) This is because—given the current state of the nonlinear art in the social sciences—virtually everyone who does this type of research is a pathbreaker, and it will be necessary to explain to others exactly what was done and how it was done. My own experience indicates that external subroutines make the body of the program seem mysterious to newcomers, especially graduate students who want to begin their own nonlinear investigations. I have found that students follow the methods more easily if every calculation is visible and each subroutine clearly labeled in the program.

The estimation program is almost entirely ready to use. Two basic requirements are that users need to place their own model into the code in the section labeled **eqs:** (beginning with line 141). It is also necessary to specify how many parameters are being estimated (variable **pnum**, see line 61). Furthermore, users need to specify the equation that holds each parameter so that the partials can be calculated (vector **parmeq**, line 63). The program is currently set up to work with individual-level survey data. Using aggregate data would require the user to weight each case by population size, and as per my annotations.

In my own experience, I have found that nonlinearities often emerge as contextually dependent phenomena. For example, things may seem fairly linear for a nation, but for certain subgroups of the population, nonlinearities can be extreme. Thus, it is often very necessary to condition the parameter estimates. This is done by writing each parameter as a linear function of a conditioning variable. If we call this new variable *CONDITION*, then we write each parameter similarly as $p = p_0 + p_1(CONDITION)$. In the following code, this is operationalized by writing **(pp[1] + (p[1]#d1))** for the parameter of interest (here the first parameter). This is done in the actual model equations (section **eqs:**). The variable **d1** is the conditioning variable. The code is written to estimate the parameter vector **p**. Thus, when unconditioned estimates are desired, the code sets the vector **pp** = 0 and **d1** = 1. When conditioned estimates are required, the code sets **pp** equal to the unconditioned estimates and **d1** equal to the conditioning variable.

As a last bit of practical advice, researchers should be aware that the coefficients of the higher degree terms in polynomial catastrophe models are sometimes difficult to estimate using simply randomly chosen initial guesses. It is often best to try a variety of initial parameter guesses that are integer values beginning with the value one. The problem is that models with higher-degree terms often have a large number of local minima of the residual sums of squares. Thus, some type of grid procedure for these particular coefficients is sometimes needed in addition to the normal random guesses for the other parameters. One suggestion is to begin by trying integer values bracketing zero for these coefficients.

The program is quite general, and other nonlinear models—not just catastrophe models—can be estimated using it. Indeed, the only real difference between estimating a catastrophe model and another type of nonlinear model is the need to be systematic rather than random with the choice of initial guesses for the parameter values of the higher-order terms of catastrophe models.

The Estimation Program

```
title 'nonlinear least-squares estimation procedure';
proc iml;
reset;
use traject;*This is the data set containing the original variables;
read all;*This reads all the variables into the separate vectors.
Vector
      and variable names are the same;
show names;*This prints the names of the vectors containing data;
```

```
*In this program, there are six state variables;
*They are r, rnext, c, cnext, l, lnext;
*Also, population and conditioning variables may be needed;
eligave=(1);*For individual level data, there is no weight per case.
      Thus, this is set equal to 1. Users of aggregate data should
change
      this to population size, or some other weight;
eligtot=sum(eligave);*Sums up the total population. In this case, it is
      simply the total number of people in the survey;
weight=eligave/eligtot;*Useful with aggregate data;
*totpop=eligtot;*Useful with aggregate data;
totpop=nrow(r);*Used for survey data;
*The next eight lines are needed for calculating the TSS;
rsqdev=eligave#((rnext-r)##2);
rssdev=rsqdev(|+,|);
lsqdev=eligave#((lnext-l)##2);
lssdev=lsqdev(|+,|);
csqdev=eligave#((cnext-c)##2);
cssdev=csqdev(|+,|);
nsqdev=eligave#((nnext-n)##2);
nssdev=nsqdev(|+,|);
r0=r;c0=c;l0=l;n0=n;*To remember the initial conditions;
cases=nrow(r);*This counts the number of cases;
print cases rssdev lssdev cssdev;
zing=0;print zing;* An initiating number for the random number
generator;
x1=uniform(zing);print x1;*The first random number;
start;
****initial parameter guesses************************;
mag1=0.50;* Used to adjust the magnitude of the initial guesses;
pp=j(30,1,0);* Creates a vector to store the parameters;
do m88=1 to 30;
      pp[m88]=uniform(x1);* Guessing strategy one;
      *pp[m88]=(2#(uniform(x1)))-1;* Guessing strategy two;
      pp[m88]=mag1#pp[m88];* The final adjustment;
end;
*** Put unconditioned parameter estimates here;* Do this only after
obtaining unconditioned estimates. Otherwise ignore the next
      three lines;
*pp[1] =0.0;*pp[2] =0.0;*pp[3] =0.0;*pp[4] =0.0;*pp[5] =0.0;*pp[6]
=0.0;
*pp[7] = 0.0;*pp[8] =0.0;*pp[9]
=0.0;*pp[10]=0.0;*pp[11]=0.0;*p[12]=0.0;
*print pp;
**********************************************************;
*Some necessary and useful numbers;
```

```
Y99 = 1;gcount=0;
i = 0.00001 ;* Used to compute the partial derivatives of the fitting surface;
h = 0.1 ;* Step size for the Runge-Kutta;
z1=j(30,1, 0.001);*Creates a vector to move the parameter estimates in
      the improving direction. This is a step size that is adjustable
      for each parameter;
p=j(30,1,0);* Another vector for storing the parameter estimates;
parmfit=j(30,1,0);*Needed for the partials;
csp=j(30,1,0);*Needed for the chi-squared tests on the parameters;
sim=j(30,1,0);*Needed for the Simon F tests (see Brown 1995);
chisq=j(30,1,0);*Needed for the chi-squared tests;
*pnum is the total number of parameters in the model;
pnum=12;*Change this as needed;
* parmeq identifies which equation each parameter is in;
parmeq={1,1,2,2,3,3,3,3,3,3,3,3,0,0,0,0,0,0,0,0,0,0,0,0,0,0,0,0,0,0};
*Initializing various counters and switches;
testlist=0;
itcount=0 ;
te=0;
te2=0;
timeup=0;
e1= 0 ;
dp1= 0 ;
*Make typedata=1 for proportions and 2 for individuals;
typedata=2;
D3 = 1 ;
***** If d3=1 you get the unconditional model *******;
***** If d3=0 you get the conditioned model *******;
if d3 = 1 then do;
      d1 = 1 ;
      p=pp;
      do m88=1 to 30;
            pp[m88]=0;
            end;
end;
if d3=0 then do;
d1=context;*This is your conditioning variable;
*Comment out the next statement if you want to use random
conditioned initial guesses;
goto norand2;
mag2=0.30;
do m88=1 to 30;
      *p[m88]=uniform(x1);*One strategy for 0 to 1 values;
      p[m88]=(2#(uniform(x1)))-1;*Another strategy for -1 to 1 values;
```

```
     p[m88]=mag2#p[m88];*Magnitude adjusting;
end;
norand2:
end;
*ESTIMATE is the main body of the program;
*The next line directs the code to that section;
goto estimate;
************************************************************;
**** The following subroutine is a fourth-order;
**** Runge-Kutta algorithm *****;
************************************************************;
modelfit:
if dp1=1 then do;*A conditional printing switch;
print 'modelfit beginning';
end;
r=r0;c=c0;l=l0;*Starting things out with the initial conditions;
do u1=1 to 10;*Iterating the Runge-Kutta ten times;
time=u1;
if (u1=1) then do;timeone=1;end;
m2=r;m1=c;m3=l;
link eqs;*EQS holds the model equations;
rk1=h#drdt;ck1=h#dcdt;lk1=h#dldt;
m2=r+(.5#rk1);m1=c+(.5#ck1);m3=l+(.5#lk1);
link eqs;
rk2=h#drdt;ck2=h#dcdt;lk2=h#dldt;
m2=r+(.5#rk2);m1=c+(.5#ck2);m3=l+(.5#lk2);
link eqs;
rk3=h#drdt;ck3=h#dcdt;lk3=h#dldt;
m2=r + rk3;m1=c + ck3;m3=l + lk3;
link eqs;
rk4=h#drdt;ck4=h#dcdt;lk4=h#dldt;
rnew=r+((1/6)#(rk1+(2#rk2)+(2#rk3)+rk4));
cnew=c+((1/6)#(ck1+(2#ck2)+(2#ck3)+ck4));
lnew=l+((1/6)#(lk1+(2#lk2)+(2#lk3)+lk4));
r=rnew;c=cnew;l=lnew;
if e1=1 then link printraj;
end;
if (typedata=1) then link propdata;
if (typedata=2) then link indidata;
link compile1;
link rsq;*RSQ is the fitting subroutine;
If dp1=1 then do;
print 'modelfit ending';
end;
return;
************************************************************;
***** Here is where the actual model **;
```

```
***** equations are placed. ****;
eqs:
*A three equation nonlinear model that I once tried;
* the variables are m1(c), m2(r), and m3(l);
link special1;
drdt = (pp[1]+(p[1]#d1))+((pp[2]+(p[2]#d1))#m2);
dcdt = (pp[3]+(p[3]#d1))+((pp[4]+(p[4]#d1))#m1);
dldt = (pp[9]+(p[9]#d1))#((((pp[5]+(p[5]#d1))
      #((pp[10]+(p[10]#d1))+w2)
      #(1-(pp[11]+(p[11]#d1))#w2)#(m3-(pp[6]+(p[6]#d1))))
      -((pp[7]+(p[7]#d1))
      #((m3-(pp[6]+(p[6]#d1)))##3))
      +((pp[8]+(p[8]#d1))#(w1+(pp[12]+(p[12]#d1))))) - m3);
return;
*************************************************************;
*Computes two variables that are needed in the model;
special1:
w1=(1+m1-m2)/2;
w2=(m2+m1)/2;
return;
*************************************************************;
printraj:* Prints the variable trajectories;
print 'trajectory for predicted r, c, and l';
r1=eligave#r;c1=eligave#c;l1=eligave#l;
traject=r1||(c1||l1);
traj=traject(|+,|);
trajmean=traj/totpop;
print trajmean;
return;
*************************************************************;
*** Computes predicted values for use in chi-square routine;
propdata:
r2=eligave#r;c2=eligave#c;l2=eligave#l;n2=eligave#n;
pred=r2||(c2||(l2||n2));
predic=pred(|+,|);
predict=predic/totpop;
if dp1=1 then do;
print 'predicted rightist, centrist, leftist, and nonvoter means';
print predict;
end;
return;
*************************************************************;
indidata:
number=nrow(r);
rpredvot=j(number,1,0);
lpredvot=j(number,1,0);
```

```
count=1;
do until (countnumber);
if (l[count] <=r[count]) then rprevdvot[count]=1;
      else rpredvot[count]=0;
if (l[count] > r[count]) then lpredvot[count]=1;
      else lpredvot[count]=0;
count=count+1;
end;
rvotes=sum(rpredvot);
lvotes=sum(lpredvot);
predic=(rvotes||lvotes);
predict=(rvotes||lvotes);
if dp1=1 then do;
print 'predicted number of rvotes and lvotes';
print predict;
end;
return;
**************************************************************;
****** The following are the subroutines
****** are for the fits and for saving
****** the best parameter estimates;
**************************************************************;
compile1:
if timeone=1 then do;
resr=0#pop;resc=0#pop;resl=0#pop; timeone=0;end;
resr=(pop#((rnext-rnew)##2));
resc=(pop#((cnext-cnew)##2));
resl=(pop#((lnext-lnew)##2));
return;
**************************************************************;
rsq:
residr=sum(resr);
residl=sum(resl);
residc=sum(resc);
rrsquare=1-(residr/rssdev);
lrsquare=1-(residl/lssdev);
crsquare=1-(residc/cssdev);
rsquare=(rrsquare+crsquare+lrsquare)/3;*The system average fit;
if dp1=1 then do;
print 'rrsquare crsquare lrsquare rsquare';
rsqfits=rrsquare||(crsquare||(lrsquare||rsquare));
print rsqfits;
end;
return;
**************************************************************;
bestpar:
```

```
bestp=p;
bestrsq = rsquare;*This is the average fit for all equations;
bestrrsq = rrsquare;
bestcrsq = crsquare;
bestlrsq = lrsquare;
parms = p;
if dp1 = 1 then do;
print , parms;
end;
return;
************************************************************;
*This computes the partial derivatives of the fitting surface;
surface:
*dp1 = 0 ;
print 'surface' 'beginning';
print 'surface' 'beginning';
do m88=1 to pnum;
p[m88]=p[m88] - i;
link modelfit;
link putfit;
p[m88]=p[m88] + (2#i);
link modelfit;
if (parmeq[m88]=1) then do;parmfit[m88]=(rrsquare - fit2);end;
if (parmeq[m88]=2) then do;parmfit[m88]=(crsquare - fit3);end;
if (parmeq[m88]=3) then do;parmfit[m88]=(lrsquare - fit4);end;
p[m88]=p[m88]-i;
end;
partials = parmfit / ( 2 # I );
test = ssq(partials);
testlist=(testlist//test);
print test, p, partials;
link modelfit;
print 'rrsquare' 'crsquare' 'lrsquare' 'rsquare';
rsqfits=rrsquare||(crsquare||(lrsquare||rsquare));
print , rsqfits;
print 'surface' 'ending';
dp1 = 0 ;
return;
************************************************************;
putfit:
fit1 = rsquare;
fit2 = rrsquare;
fit3 = crsquare;
fit4 = lrsquare;
return;
************************************************************;
```

```
putpred:
predic1 = predic;
preds1 = rsquare;
return;
************************************************************;
csfits:
if (typedata=1) then do;
predic1=predic1+1;predic=predic+1;end;
chsqall1 = ((predic - predic1)##2) / abs(predic1);
chsqall = sum(chsqall1);
simall = preds1 - rsquare;
print chsqall simall;
return;
***********************************************************;
*The following is the main body of the program;
estimate:
dp1 = 1;te = 1;e1 = 0;
print p;
link modelfit;te = 0;e1 = 0;
link bestpar;
dp1 = 0 ;
do y99 = 1 To 30;
zcount = 0 ;
itcount = 0 ;
link surface;
z2 = z1;
*print test, z2;
print 'estimate' 'beginning';
e1 = 0 ;
link modelfit;
e1 = 0 ;
begin:
fit1 = rsquare;
newparm = p + ( partials # z2 );*Improving the estimates;
p = newparm;
link modelfit;
if rsquare > fit1 then do;
link bestpar;
end;
itcount = itcount + 1 ;
if rsquare > fit1 then goto begin;
if zcount<2 then do; print itcount;end;
itcount = 0 ;
zcount = zcount + 1 ;
*print , zcount;
z2 = z2 / 10 ;
```

```
if zcount > 4 then goto jump1;
p=bestp;
goto begin;
jump1:
p=bestp;
end;
print 'estimate' 'ending';
te = 1 ;link modelfit;te = 0;
goto chisquar;
**************************************************************;
*Calculating the chi-squared statistics;
chisquar:
te2=1;
do m88=1 to pnum;
csp[m88] = p[m88];link modelfit;link putpred;
p[m88] = 0 ;link modelfit;link csfits;
chisq[m88] = chsqall;sim[m88]=simall;p[m88] = csp[m88];
end;
**** Now the overall model chi square;
csp = p;link modelfit;link putpred;
do m88=1 to 11;
      p[m88] = 0 ;end;
link modelfit;link csfits;
csmodel = chsqall;simmodel=simall;p = csp;
**************************************************************;
**** Preparation for output ****************************;
**************************************************************;
dp1 = 1 ;
link modelfit;
dp1 = 0 ;
allfit1 = ( rrsquare || (crsquare||(lrsquare||rsquare)));
syspar = bestp;
syschi = chisq;
syssim = sim;
allfits = shape( allfit1 ,0 , 1 );
syschisq = shape( syschi ,0 , 1 );
syssims = shape( syssim ,0 , 1 );
sysparms = shape( syspar ,0 , 1 );
sysest = (sysparms||(syschisq||syssims));
print , c44 , sysparms , syschisq , sysest , testlist;
print csmodel, simmodel;
dp1=1;
link modelfit;
dp1=0;
stuff={parms chisq simon};
row = {'p1' 'p2' 'p3' 'p4' 'p5' 'p6' 'p7' 'p8' 'p9' 'p10'
```

```
            'p11' 'p12' 'p13' 'p14' 'p15' 'p16' 'p17' 'p18' 'p19' 'p20'
            'p21' 'p22' 'p23' 'p24' 'p25' 'p26' 'p27' 'p28' 'p29' 'p30'};
create betas from sysest (|colname=stuff rowname=row|);
append from sysest (|rowname=row|);
close betas;
fit={'fits'};
roww = {'eq1' 'eq2' 'eq3' 'ave'};
create fits from allfits (|colname=fit rowname=roww|);
append from allfits (|rowname=roww|);
close fits;
finish;run;
quit;
proc print data=betas;var row parms chisq simon;
proc print data=fits;var roww fits;
```

Notes

Chapter 1

1. For a more extended discussion of this subject, see Brown 1991, chapter 3.

2. This specification unrealistically assumes an unlimited supply of food for the prey. Many modifications of this model have since been developed that add a substantial degree of structural realism to the original specification. In particular, see discussions by Danby (1985).

Chapter 2

1. Indeed, this author has encountered many instances in which the estimation of a continuous time nonlinear model revealed strong support for the right-hand side algebraic structure of the model while the exact same algebra produced very disappointing results with a discrete dependent variable and a regression estimation. The cause of the discrete failure is not the algebra on the right-hand side, but the way change is measured on the left-hand side.

2. On the opposite end of the spectrum, linear regression models can be viewed as "general" to an excess. Indeed, it is quite surprising that so little debate has occurred in the social sciences regarding the seemingly countless number of empirical studies that employ basically the same algebraic form across dramatically different real situations. Thus, we have an anthropologist studying a tribal economy, a psychologist examining cognitive structures of experimental subjects, and a political scientist researching voting behavior all using the same linear algebra (i.e., the same model). It is natural to wonder how distorted social theory must become when it is routinely boiled down into a linear algebra due to practical estimation requirements.

3. As compared with, say, the nonlinear beauty of the shape of a leaf, or its tumbling fall.

Chapter 3

1. There have, of course, been other attempts to investigate the interaction between individual rationality autonomy and the influence of context, and one highly interesting line of research has been to question the degree of political freedom that individuals have within the context of strong cultural norms (Gibson 1992).

2. From an empirical point of view, other polynomial formulations (including higher degree structures) were attempted but found to be no more advantageous than the cubic form developed here. Squaring C did not yield satisfactory empirical results.

3. Some readers may feel that the appearance of a catastrophe structure is merely a product of the polynomial algebra, in the sense that if one removes the cubed terms the catastrophe vanishes. Remember, however, that normally there must always be an isomorphic correspondence between a model's algebra and a substantive understanding of a social phenomenon. In the current setting, one cannot simply remove a part of the algebra without changing the social theory that I have developed here. The theory and the algebra are one and the same. Thus, the catastrophe structure is not a function of the algebra. Rather, the algebra is a requirement of my theoretical development. If one challenges my specification of this process, one must first challenge my theory that led me to anticipate bimodal expectations with regard to change in voter feelings.

4. All equilibria surfaces presented here are computed using numerical applications of Newton's method for finding roots of equations.

Chapter 4

1. Thus, this analysis does not characterize landslide elections as structurally different from other elections, although they may indeed be different in this regard. Rather, the focus here is on the description of the highly complex nature of the contextually dependent aggregate electoral topology of one landslide, from which future comparisons may be made to other elections.

2. Some readers may wish that the current study also include a truly comparative analysis of other landslides from an aggregate perspective. Such an analysis is entirely possible, and much work along these lines has already been accomplished by the author. Moreover, the presentation of additional results is planned for the future. But normal limitations of space require that the focus of the current analysis remain on one landslide, and the choice is made to examine the largest-magnitude landslide in this century. In some respects, the current investigation can serve as a baseline analysis for the examination of other similarly dramatic electoral phenomena.

3. See White 1965, Shadegg 1965, and Cosman 1966 for detailed substantive analyses of the 1964 election.

4. A random mixing assumption between the two populations is not necessary, although it may be heuristically useful in introducing the model (see Mesterton-Gibbons 1989).

5. Some readers may wonder if the inclusion of the limit terms [i.e., $(1 - D)D$] confuses the interpretation of the other parts of the model. This does not occur since each of the terms participates within the model's overall structure in a unique way. For example, the limit terms turn the entire derivative off at the margins of substantive plausibility. None of the other terms act on the entire model in this way. For example, the term jD has no effect on the term vN, or the term $q(D/R)$ when it is multiplied by the 1 in the first term of the model. In general, such models as presented in equation 4.6 are quite sensitive to the algebraic uniqueness of the terms, and thus are quite successful in differentiating the relative importance of these terms (see Mesterton-Gibbons 1989).

6. The model, as expressed in equations 4.6 and 4.7, is a deterministic characterization of social change. The only difference between a deterministic model and a probabilistic model is the existence of distributional assumptions with regard to the predicted state variables (see Brown 1995, chap. 1). Indeed, the equation of a line is a deterministic model if one makes no distributional assumptions with regard to the dependent variable, as is characteristic of ordinary least squares. In practice, probabilistic models are much more limited than deterministic models in terms of algebraic flexibility since reasonable distributional assumptions tend to fall apart when the models depart dramatically from a strictly linear form. This occurs, for example, when a model contains a nonlinearity with regard to the parameters (see Judge et al. 1982, 633–63). This distinction between probabilistic and deterministic model forms affects both the optimization procedure used to estimate the parameters as well as the choice of statistical tests used to evaluate the significance of the estimated parameters. For a thorough comparison of both types of model structures, see Mesterton-Gibbons 1989.

7. All of the data utilized in this analysis were supplied by the Inter-University Consortium for Political and Social Research. Of course, I alone am responsible for all of the interpretations presented here.

8. See Brown 1991, chapter 3 for a more generalized discussion of this topic.

9. The term *Simon-effects* references early formal investigations of social systems by Herbert A. Simon (1957) in which algebraic structures were specifically tied to aggregate, and thus social, human experience (see also Brown 1988, 1991). The numbers themselves express the absolute value of the change in the model's predicted level of partisan support, measured as a proportion of the eligible electorate, that occurs when a given parameter is set equal to zero as compared with that obtained using its optimized value. The magnitudes of the Simon-effects can be compared across parameters within one partisan model to determine the relative impact of each of the parameters in affecting change in partisan mobilization. The Simon-effects cannot, however, be compared across equations. For example, in table 4.2, parameters p and s represent the two acceleration influences that are caused by the Republican campaign's momentum. (Substantively, one should recall that momentum can be positive or negative, depending on whether a party is increasingly winning or losing support, respectively). The approximately equal magnitude of the two estimates for the parameters p and s suggests that both acceleration influences played relatively equal parts in the Republican loss in 1964. On the other hand, with respect to Democratic voting, the estimate for parameter j is more than twice the magnitude of the estimate for parameter y, suggesting that the accelerated momentum of the Democratic campaign in areas outside the Deep South was more highly structured by the proportional local strength of Democratic support rather than by the impact of Democratic voters interacting with other Democratic voters (i.e., the bunching effect).

10. The initial conditions of the actual data are not used in the figure, since there are too many data points to make a clearly readable graph.

11. The analysis of survey data presented here is methodologically straightforward. Much more complex analyses of these data were conducted during the course of these investigations, including multiequation logistic regression structures. However, none of these analyses produced results that added significantly to the interpretive guidelines drawn from the t-tests presented in table 4.4.

Chapter 5

1. Changing the measure of deinstitutionalization to reflect changes in vote mobilization rather than vote share would lead to an alternate, but nonetheless interesting, set of analyses compared with those presented here. Such analyses have, in fact, been completed, and the intention is to present the results in another setting. Moreover, one need not think that deinstitutionalization based on vote share on the national level only influences partisan fragmentation. People observe the partisan balance within their local environment, and system changes with regard to partisan fragmentation are likely to occur on the local level as well. See Brown 1991 for a review of findings and literature relating to the influence of localized contextual effects on macropolitical systems.

2. The second election in 1932 can be thought of as sort of an "aftershock" with regard to a national awakening to the new political realities. Moreover, most scholars consider the 1933 election to be less representative of the national mood due to the heavy-handed tactics used during the campaign by the Nazis.

3. Note that the interactions can be both direct and indirect. A typical example of a direct interaction is a conversation between two or more people. Indirect interactions can take many forms, including passive observations of local individual and group activities. The micromechanisms by which such interactions influence individual and group behaviors and attitudes is still an area of ongoing research among a number of scholars. However, that the influences exist is beyond doubt. In particular, see Huckfeldt and Sprague 1988, 1987, Molotch and Boden 1985, Garfinkel 1967, Gurwitsch 1962, Blau 1977, Blum 1985, and Simmel 1955.

4. Indeed, the relationship between realistic limits and theoretical limits is quite murky in the literature on dynamics, both in the social sciences and occasionally elsewhere. The problem rests in the difficulty that is often encountered in estimating the realistic limits. The problem is often addressed by simply inserting the theoretical limit into the model and forgoing the matter of estimating the realistic limit. Thus, the current research should be of general interest to theorists in that it allows researchers a chance to see, in this one example, the proximity of these limits.

5. It is worth pointing out that the estimation of such systems is entirely nontrivial, and it is possible only due to the recent advances in computing technology. The estimates presented here in tables 5.3 and 5.4 took the author nearly two years to obtain by running an estimation program on an IBM 3090 supercomputer for literally thousands of hours of real cpu time.

6. The data used to construct this figure were generated from a program written by the author in Pascal. The visual image was then constructed from these data using imaging software written by the National Center for Supercomputing Applications (NCSA) Software Development Group at the University of Illinois at Urbana-Champaign.

Chapter 6

1. Indeed, the collapse of the former Communist governments of Eastern Europe and the former Soviet Union led to near immediate revelations of how deep

and widespread such damage can be. (See *New York Times*, 13 May 1992, A1 and A4[N].)

2. See Mesterton-Gibbons 1989 for a more complete overview of such guidelines.

3. In all of the numerical results presented in this paper, the half-life for the system's environmental damage was fixed at six years.

4. Recent anecdotal evidence nicely illustrates how sitting presidents can and do change regulatory environmental policy to achieve electoral benefit with regard to current constituents. The following list of articles and editorials, all drawn from the national edition of the *New York Times* during the 1992 presidential primary season, describe how President Bush made an effort to reduce, waive, or eliminate environmental regulations that adversely affected potential or current supporters of his campaign. This partial listing is *not* meant to imply that President Bush is unique in any way in this regard. The argument made here is that this is a natural occurrence of partisan politics in the United States in which environmental concerns fall victim to the same constituency pressures as any other issue or set of issues. The titles and dates of the articles are listed. "Environment Laws are Eased by Bush as Election Nears," 20 May 1992, A1 and A10[N]; "Bush to Relax 1990 Rule On Air Pollution Notices," 18 May 1992, A9[N], "2 Admit E.P.A. Violated Hazardous Waste Law," 8 May 1992, A10[N]; "Administration Tries to Limit Rule Used to Halt Logging of National Forests," 28 April 1992, A7[N]; "E.P.A. Head Allows Project On a Lake Michigan Marsh," 9 May 1992, A7[N]; "White House and Congress Face Showdown on National Forests," 12 May 1992, A8[N]; "Showdown on Endangered Species," 11 May 1992, A15[N]; "Icy Words on Global Warming," 30 March 1992, A14[N]; "Freezing Government," 4 May 1992, A14[N]; "The Environmental Pollution President," 29 April 1992, A15[N].

5. I am indebted to G. Robert Boynton for suggesting this second justification for parameter p.

6. This approach to the numerical evaluation of such systems is described more fully in Kocak 1989 as well as Mesterton-Gibbons 1989.

7. Throughout this analysis, all numerical solutions to the differential equations were obtained using a fourth-order Runge-Kutta algorithm. Moreover, all of the results presented here were obtained using software written by the author in Pascal for a Macintosh Quadra 900. The computer was made available to the author by James Johnson, Vice Provost of Emory University's Information Technology Division.

8. In figures 6.1 through 6.4, the parameter r is given a value of one.

9. Of course, actual partisan changes would not be as evenly spaced. Holding the spacing constant here, however, acts to control the structure of this new input, thereby clearly identifying its effect on the overall system.

10. By way of casual observation, this scenario indeed seems a partial contributor to what may have happened to the Great Lakes of North America. The public's response to the increasing levels of pollution was very slow, probably due to the indirect ways that lake pollution affected peoples' lives.

11. For example, the Bush administration passed a flurry of regulations regarding the environment in the final days of its tenure (the *New York Times*, 16 January 1993, A1[N]). Similarly, the Reagan administration engineered a dramatic decline in funding

for the Environmental Protection Agency during President Reagan's first term in office (Wood 1988).

12. Figures 6.6 and 6.7 were constructed using imaging software provided by the National Center for Supercomputing Applications at the University of Illinois at Urbana-Champaign.

13. The zero and negative frequency elements are suppressed in this computation of total power. A useful introduction to this type of analysis can be found in Press, Flannery, Teukolsky, and Vetterling 1989.

14. Chaos, classically defined, indicates extreme longitudinal complexity, to the point of mimicking random variation (see Brown 1995).

References

Abraham, Ralph H., and Christopher D. Shaw. 1992. *Dynamics: The Geometry of Behavior*. 2d edition. Redwood City, Calif.: Addison-Wesley.

Andersen, Kristi. 1979. *The Creation of a Democratic Majority 1928–1936*. Chicago: University of Chicago Press.

Arendt, Hannah. 1961. *Between Past and Future*. New York: Faber and Faber.

Asher, Herbert B. 1988. *Presidential Elections and American Politics: Voters, Candidates, and Campaigns since 1952*. Chicago: Dorsey Press.

Bard, Y. 1974. *Nonlinear Parameter Estimation*. New York: Academic Press.

Barrilleaux, Charles J. 1986. A Dynamic Model of Partisan Competition in the American States. *American Journal of Political Science* 30:822–40.

Beck, Paul Allen. 1974. Environment and Party: The Impact of Political and Demographic County Characteristics on Party Behavior. *American Political Science Review* 68:1229–44.

Berelson, Bernard R., Paul F. Lazarsfeld, and William N. McPhee. 1954. *Voting: A Study of Opinion Formation in a Presidential Campaign*. Chicago: University of Chicago Press.

Black, Earl, and Merle Black. 1987. *Politics and Society in the South*. Cambridge, Mass.: Harvard University Press.

Blau, Peter M. 1977. *Inequality and Heterogeneity*. New York: Free Press.

Blum, Terry C. 1985. Structural Constraints on Interpersonal Relations: A Test of Blau's Macrosociological Theory. *American Journal of Sociology* 91:511–21.

Brown, Courtney. 1982. The Nazi Vote: A National Ecological Study. *American Political Science Review* 76:285–302.

Brown, Courtney. 1987. Voter Mobilization and Party Competition in a Volatile Electorate. *American Sociological Review* 52:59–72.

Brown, Courtney. 1988. Mass Dynamics of U.S. Presidential Competitions, 1928–36. *American Political Science Review* 82:1153–81.

Brown, Courtney. 1991. *Ballots of Tumult: A Portrait of Volatility in American Voting*. Ann Arbor, Mich.: Univeristy of Michigan Press.

Brown, Courtney. 1993. Nonlinear Transformation in a Landslide: Johnson and Goldwater in 1964. *American Journal of Political Science*. 37:582–609.

Brown, Courtney. 1994. Politics and the Environment: Nonlinear Instabilities Dominate. *American Political Science Review* 88:292–303.

Brown, Courtney. 1995. *Chaos and Catastrophe Theories: Nonlinear Modeling in the Social Sciences*. In the Quantitative Applications in the Social Sciences series. Beverly Hills, Calif.: Sage.

Burnham, Walter Dean. 1970. *Critical Elections and the Mainsprings of American Politics*. New York: W.W. Norton.

Campbell, Angus, Philip E. Converse, Warren Miller, and Donald E. Stokes. 1960. *The American Voter*. New York: Wiley.

Canovan, Margaret. 1974. *The Political Thought of Hannah Arendt*. New York: Harcourt Brace Jovanovich.

Coleman, James S. 1964. *Introduction to Mathematical Sociology*. New York: Free Press.

Coleman, James S. 1981. *Longitudinal Data Analysis*. New York: Basic Books.

Converse, Philip. 1975. Public Opinion and Voting Behavior. In *The Handbook of Political Science*, ed. Fred A. Greenstein and Nelson Polsby. Reading, Mass: Addison-Wesley.

Cortez, Fernando, Adam Przeworski, and John Sprague. 1974. *Systems Analysis for Social Scientists*. New York: Wiley.

Cosman, Bernard. 1966. *Five States for Goldwater: Continuity and Change in Southern Presidential Voting Patterns*. Birmingham, Ala.: University of Alabama Press.

Crosby, Robert W. 1987. Toward a Classification of Complex Systems. *European Journal of Operational Research* 30:291–93.

Danby, J. M. A. 1985. *Computing Applications to Differential Equations: Modelling in the Physical and Social Sciences*. Reston, Va.: Reston Publishing Company.

Darmstadter, Joel. 1991. The Economic Cost of CO_2 Mitigation: A Review of Estimates for Selected World Regions. Discussion Paper ENR91-06, Resources for the Future, Washington, D.C.

Dennis, John E., and Robert B. Schnabel. 1983. *Numerical Methods for Unconstrained Optimization and Nonlinear Equations*. Englewood Cliffs, N.J.: Prentice-Hall.

Downs, Anthony. 1957. *An Economic Theory of Democracy*. New York: Harper & Row.

Energy Modeling Forum. 1992. Global Climate Change: Energy Sector Impacts of Greenhouse Gas Emission Control Strategies. Energy Modeling Forum Report No. 12, Stanford University.

Erikson, Robert S., and Kent L. Tedin. 1981. The 1928–1936 Partisan Realignment: The Case for the Conversion Hypothesis. *American Political Science Review* 75:951–62.

Eulau, Heinz, and Lawrence Rothenberg. 1986. Life Space and Social Networks as Political Contexts. *Political Behavior* 8:130-57.

Garfinkel, Harold. 1967. *Studies in Ethnomethodology*. Englewood Cliffs, N.J.: Prentice-Hall.

Gibson, James L. 1992. The Political Consequences of Intolerance: Cultural Conformity and Political Freedom. *American Political Science Review* 86:338–56.

Gillespie, John V., Dina A. Zinnes, Philip A. Schrodt, G. S. Tahim, and R. Michael Rubinson. 1977. An Optimal Control Model of Arms Races. *American Political Science Review* 71:226–44.

Goldberg, Samuel. 1958. *Introduction to Difference Equations*. New York: John Wiley and Sons.

Grofman, Bernard. 1987. Models of Voting. *Research in Micropolitics* 2:31–61.

Gurwitsch, Aron. 1962. The Common-sense World as Social Reality—A Discourse on Alfred Schultz. *Social Research* 29:50–72.

Haberman, Richard. 1977. *Mathematical Models: Mechanical Vibrations, Population Dynamics, and Traffic Flow*. Englewood Cliffs, N.J.: Prentice-Hall.

Hamilton, Richard F. 1982. *Who Voted for Hitler?* Princeton: Princeton University Press.

Hamming, Richard W. 1971. *Introduction to Applied Numerical Analysis*. New York: McGraw-Hill.

Hamming, Richard W. 1973. *Numerical Methods for Scientists and Engineers*. 2d ed. New York: McGraw-Hill.

Harris, Richard A., and Sidney M. Milkis. 1989. *The Politics of Regulatory Change: A Tale of Two Agencies*. New York: Oxford University Press.

Hibbs, Douglas A., Jr. 1977. Political Parties and Macroeconomic Policy. *American Political Science Review* 71:1467–87.

Hirsch, Morris W., and Stephen Smale. 1974. *Differential Equations, Dynamical Systems, and Linear Algebra*. New York: Academic Press.

Huckfeldt, R. Robert. 1980. Variable Responses to Neighborhood Social Contexts: Assimilation, Conflict, and Tipping Points. *Political Behavior* 2:231–57.

Huckfeldt, R. Robert. 1983. The Social Context of Political Change: Durability, Volatility, and Social Influence. *American Political Science Review* 77:929–44.

Huckfeldt, R. Robert, and Carol Weitzel Kohfeld. 1989. *Race and the Decline of Class in American Politics*. Urbana: University of Illinois Press.

Huckfeldt, R. Robert, C. W. Kohfeld, and Thomas W. Likens. 1982. *Dynamic Modeling: An Introduction*. Beverly Hills, Calif.: Sage.

Huckfeldt, R. Robert, Erix Plutzer, and John Sprague. 1993. Alternative Contexts of Political Behavior: Churches, Neighborhoods, and Individuals. *The Journal of Politics* 55:365–81.

Huckfeldt, R. Robert, and John Sprague. 1987. Networks in Context: The Social Flow of Political Information. *American Political Science Review* 81:1197–1216.

Huckfeldt, R. Robert, and John Sprague. 1988. Choice, Social Structure, and Political Information: The Informational Coersion of Minorities. *American Journal of Political Science* 32:467–82.

Huckfeldt, R. Robert, and John Sprague. 1992. Political Parties and Electoral Mobilization: Political Structure, Social Structure, and the Party Canvass. *American Political Science Review* 86:70–86.

Huntington, Samuel P. 1968. *Political Order in Changing Societies*. New Haven, Conn.: Yale University Press.

Judge, George G., R. Carter Hill, William Griffiths, Helmut Lutkepohl, and Tsoung-Chau Lee. 1982. *Introduction to the Theory and Practice of Econometrics*. New York: Wiley.

Kahneman, Daniel, and Amos Tversky. 1982. The Psychology of Preferences. *Scientific American*. 34(January):160–73.

Kelley, Stanley, Jr. 1983. *Interpreting Elections*. Princeton, N. J.: Princeton University Press.

Key, V. O. 1955. A Theory of Critical Elections. *Journal of Politics* 17:3–18.

Key, V. O. 1959. Secular Realignment and the Party System. *Journal of Politics* 21:198–210.

Kocak, Huseyin. 1989. *Differential and Difference Equations through Computer Experiments*. 2d ed. New York: Springer-Verlag.

Koppstein, Peter E. A. 1983. Parameterized Dynamical Systems Perspectives on Collective Action with Special References to Political Protest and Civil Violence. Ph.D. diss., Department of Political Science, Yale University.

Kramer, Gerald H. 1983. The Ecological Fallacy Revisited: Aggregate versus Individual-Level Findings on Economics and Elections and Sociotropic Voting. *American Political Science Review* 77:92–111.

Laakso, Markku, and Rein Taagepera. 1979. Effective Number of Parties: A Measure with Application to West Europe. *Comparative Political Studies* 12: 3-27.

Ladd, Everett Carll, and Charles D. Hadley. 1978. *Transformations of the American Party System: Political Coalitions from the New Deal to the 1970's*. New York: W.W. Norton.

Leslie, P.H. 1948. Some Further Notes on the Use of Matrices in Population Mathematics. *Biometrika* 35:213–45.

Lipset, Seymour Martin. 1981. *Political Man: The Social Bases of Politics*. New York: Doubleday.

Luenberger, David G. 1979. *Introduction to Dynamic Systems: Theory, Models, and Applications*. New York: Wiley.

McBurnett, Michael. 1995a. Order and Complexity in the Evolution of Public Opinion. In *Chaos Theory in the Social Sciences: Foundations and Applications*, ed. L. Douglas Kiel and Euel Elliott. Ann Arbor, Mich.: University of Michigan Press.

McBurnett, Michael. 1995b. Probing the Underlying Structure in Dynamical Systems: The Uses of Spectral Analysis. In *Chaos Theory in the Social Sciences: Foundations and Applications*, ed. L. Douglas Kiel and Euel Elliott. Ann Arbor, Mich.: University of Michigan Press.

MacKuen, Michael Bruce. 1981. Social Communication and the Mass Policy Agenda. In *More than News*, by Michael Bruce MacKuen and Steven Lane Coombs. Beverly Hills, Calif.: Sage.

MacKuen, Michael, and Courtney Brown. 1987. Political Context and Attitude Change. *American Political Science Review* 81:471–90.

McPhee, William N. 1963. *Formal Theories of Mass Behavior*. New York: Macmillan.

McPhee, William N., and Jack Ferguson. 1962. Political Immunization. In *Public Opinion and Congressional Elections*, ed. William N. McPhee and William A. Glaser, 155–79. New York: Free Press.

McPhee, William N., and Robert B. Smith. 1962. A Model for Analyzing Voting Systems. In *Public Opinion and Congressional Elections*, ed. William N. McPhee and William A. Glaser, 123–54. New York: Macmillan.

May, Robert M. 1974. *Stability and Complexity in Model Ecosystems*. Princeton, N.J.: Princeton University Press.

May, Robert M. 1976. Simple Mathematical Models with Very Complicated Dynamics. *Nature* 26:459–67.

Mesterton-Gibbons, Michael. 1989. *A Concrete Approach to Mathematical Modeling*. New York: Addison-Wesley.

Miller, James Grier. 1978. *Living Systems*. New York: McGraw-Hill.

Miller, Warren E. 1956. One-Party Politics and the Voter. *American Political Science Review* 50:707–25.

Molinar, Juan. 1991. Counting the Number of Parties: An Alternative Index. *American Political Science Review* 85:1383–91.

Molotch, Harvey L., and Deirdre Boden. 1985. Talking Social Structure: Discourse, Domination and the Watergate Hearings. *American Sociological Review* 50:273–88.

Moulton, Robert J., and Kenneth R. Richards. 1990. Costs of Sequestering Carbon Through Tree Planting and Forest Management in the United States. December, General Technical Report WO-58, Forest Service, U.S. Department of Agriculture, Washington, D.C.

Nordhaus, William D. 1991a. Economic Approaches to Greenhouse Warming. In *Global Warming: Economic Policy Responses*, ed. Rudiger Dornbusch and James M. Poterba, 33–66. Cambridge, Mass: MIT Press.

Nordhaus, William D. 1991b. The Cost of Slowing Climate Change: A Survey. *Energy Journal* 12:37–66.

Petrocik, John R. 1981. *Party Coalitions: Realignments and the Decline of the New Deal Party System*. Chicago: University of Chicago Press.

Poterba, James M. 1991. Tax Policy to Combat Global Warming: On Designing a Carbon Tax. In *Global Warming: Economic Policy Responses*, ed. Rudiger Dornbusch and James M. Poterba, 71–97. Cambridge, Mass: MIT Press.

Przeworski, Adam. 1975. Institutionalization of Voting Patterns, or is Mobilization the Source of Decay. *American Political Science Review* 69:49–67.

Przeworski, Adam, and Glaucio A. D. Soares. 1971. Theories in Search of a Curve: A Contextual Interpretation of Left Vote. *American Political Science Review* 65:51–65.

Przeworski, Adam, and John Sprague. 1986. *Paper Stones: A History of Electoral Socialism*. Chicago: University of Chicago Press.

Rae, Douglas W. 1967. *The Political Consequences of Electoral Laws*. New Haven, Conn.: Yale University Press.

Rapoport, Anatol. 1963. Mathematical Models of Social Interaction. In *Handbook of Mathematical Psychology*, ed. R. Duncan Luce, Robert R. Bush, and Eugene Galanter, 2:145–85. New York: Wiley.

Rapoport, Anatol. 1983. *Mathematical Models in the Social and Behavioral Sciences*. New York: Wiley.

Rashevsky, Nicolas. 1954. Two Models: Imitative Behavior and Distribution of Status. In *Mathematical Thinking in the Social Sciences*, ed. Paul F. Lazarsfeld. Glencoe, Ill.: Free Press.

Richards, Diana. 1992. Spatial Correlation Test for Chaotic Dynamics in Political Science. *American Journal of Political Science* 36:1047–69.

Richardson, Lewis F. 1960. *Arms and Insecurity: A Mathematical Study of the Causes and Origins of War*. Pittsburgh: Boxwood.

Rowland, C. K., S. C. Lee, and D. B. Goetze. 1990. Longitudinal and Catastrophic

Models of State Hazardous Waste Regulation. In *Economics, Institutions, and Public Policy*, ed. Allen Gitelson and Melvin Dubnick. Greenwood, Conn.: JAI Press.

Saunders, P. T. 1980. *An Introduction to Catastrophe Theory*. New York: Cambridge Univerisity Press.

Schelling, Thomas C. 1978. *Micromotives and Macrobehavior*. New York: Norton.

Schelling, Thomas C. 1992. Some Economics of Global Warming. *American Economic Review* 82:1–14.

Scherer, F. M. 1980. *Industrial Market Structure and Economic Performance*. Chicago: Rand McNally.

Shadegg, Stephen C. 1965. *What Happened to Goldwater?: The Inside Story of the 1964 Republican Campaign*. New York: Holt, Rinehart, and Winston.

Simmel, Georg. 1955. *The Web of Group Affiliations*. Glencoe, Ill.: Free Press.

Simon, Herbert A. 1957. *Models of Man: Social and Rational*. New York: Wiley.

Sprague, John. 1976. Estimating a Boudon Type Contextual Model: Some Practical and Theoretical Problems of Measurement. *Political Methodology* 3:333–53.

Sprague, John. 1980. Two Variants of Aggregation Processes and Problems in Elementary Dynamic and Contextual Causal Formulations. Political Science Paper No. 50, Department of Political Science, Washington University, St. Louis.

Sprague, John. 1981. One-Party Dominance in Legislatures. *Legislative Studies Quarterly* 6:259–85.

Stimson, James A. 1991. *Public Opinion in America: Moods, Cycles, and Swings*. Boulder, Colo.: Westview Press.

Sundquist, James L. 1983. *Dynamics of the Party System: Alignment and Realignment of Political Parties in the United States*. Washington, D.C.: Brookings Institution.

Taagepera, Rein, and Matthew Soberg Shugart. 1993. Predicting the Number of Parties: A Quantitative Model of Duverger's Mechanical Effect. *American Political Science Review* 87:455–64.

Thom, Rene. 1975. *Structural Stability and Morphogenesis; An Outline of a General Theory of Models*. Reading, Mass.: W. A. Benjamin.

Tobin, Richard. 1990. *The Expendable Future: U.S. Politics and the Protection of Biological Diversity*. Durham, N.C.: Duke University Press.

Tuma, Nancy Brandon, and Michael T. Hannan. 1984. *Social Dynamics: Models and Methods*. New York: Academic Press.

Vig, Norman J., and Michael E. Kraft. 1984. *Environmental Policy in the 1980's: Reagan's New Agenda*. Washington, D.C.: Congressional Quarterly.

Ward, Michael Don. 1984. Differential Paths to Parity: A Study of the Contemporary Arms Race. *American Political Science Review* 78:297–317.

Waterman, Richard W. 1989. *Presidential Influence and the Administrative State*. Knoxville: University of Tennessee Press.

White, Theodore Harold. 1965. *The Making of the President, 1964*. New York: Atheneum Publishers.

Wildgen, John K. 1971. The Measurement of Hyperfractionalization. *Comparative Political Studies* 4:233–43.

Wood, B. Dan. 1988. Principals, Bureaucrats, and Responsiveness in Clean Air Enforcements. *American Political Science Review* 82:213–34.

Wood, B. Dan, and Richard W. Waterman. 1991. The Dynamics of Political Control of the Bureaucracy. *American Political Science Review* 85:801–28.

Index